The Bee Book

Also by the same author:

Ideas for Interesting Gardens

To the Reader

The Chiefest cause, to read good bookes,
That moves each studious minde
Is hope, some pleasure sweet therein,
Or profit goode to finde.
Now what delight can greater be
Than secrets for to knowe,
Of Sacred Bees, the Muses Birds,
All which this booke doth shew.
And if commodity thou crave,
Learne here no little gaine
Of their most sweet and sov'raigne fruits,
With no great cost or paine,
If pleasure then, or profit may
To read induce thy minde;
In this smale Treatise choice of both,
Good Reader, thou shalt finde.

Charles Butler
The Feminine Monarchie
1609

The Bee Book

The History and Natural History of the Honeybee

Daphne More

David & Charles
Newton Abbot · London · Vancouver

ISBN 0 7153 7268 8

Set in 10 on 11 Latinesque
and printed in Great Britain
by Ebeneezer Baylis & Son Ltd
The Trinity Press
Worcester & London
for David & Charles (Publishing) Limited
Brunel House Newton Abbot Devon

Published in Canada
by Douglas David & Charles Limited
1875 Welch Street North Vancouver BC

Contents

Acknowledgements

My especial thanks are due to F. G. Vernon who took the photographs for this book except where credited otherwise, and whose wide knowledge, practical help and encouragement have been of the greatest value to me.

I wish also to thank the International Bee Research Association, and in particular Dr Eva Crane, for permission to reproduce copyright photographs of items in the IBRA Collection of Historical and Contemporary Beekeeping Material. This collection was founded in 1952 and is housed partly at the Association's headquarters at Hill House, Chalfont St Peter, Bucks, and partly at the Museum of English Rural Life at the University of Reading, Berks. A booklet about the collection by F. G. Vernon is available from IBRA. Any reader who comes across such equipment, or papers and publications relating to bygone beekeeping, is urged to contact the Association with a view to ensuring their preservation.

Drawings, except where otherwise credited, are by the author.

My thanks are due to all the people who gave me information and practical help, particularly to the Worshipful Company of Wax Chandlers; Harald Pager and his publisher, Akademische Druck-u, Verlagsanstalt, Graz, Austria; L. L. Thorne of E. C. Thorne Ltd; R. S. Pitcher of Price's Patent Candle Co; H. A. Sturdy of Poth, Hille and Co, wax-refiners; L. W. Kaye of Charles Farris Ltd, candlemakers; and staff of Madame Tussaud's, the British Museum and British Library, the Metropolitan Museum of Art, New York, and the City Museum, Winchester.

I wish it were also possible to thank all those authors from Aristotle onwards who found the honeybee as fascinating as I do, and whose works I have read with so much pleasure.

1 The Origin of Bees

Mankind has been fascinated by bees from the beginning of history. He realised that they were strange and marvellous creatures and he found no difficulty in believing the many fantastic stories which grew up over the centuries when few facts about them were known. Some of the oddest ideas concerned the origin of bees. They existed, leading their secret lives in caves and hollow trees, sometimes occupying homes of Man's providing though never really domesticated; but where had they come from in the beginning?

A Breton legend tells how the falling tears of the crucified Christ turned into bees and flew away to bring sweetness to the world. Bees, however, existed long before the first apeman staightened his back and looked from beneath jutting brows at an already-ancient world. Primeval man soon discovered that bees provided food—the main preoccupation of his existence—and he learned to rob the nests he found in the forests and to eat not only the honeycomb but also the eggs and larvae, as primitive people still do.

Most ancient races regarded bees with awe and they had a place in many pagan religions. The Rig-Veda, the Hindu holy book written in Sanskrit between 2000 and 3000 BC, often refers to them. Vishnu the Preserver, one of the great triad with Brahma and Siva, is represented as a blue bee on a lotus flower; and Kama, the Indian god of love, carries a bow, the string of which is a chain of bees.

The earliest almost universal deity was the Earth or Great-Mother, always a goddess of fertility, wildlife and agriculture although she had different names—Isis, Ceres, Cybele, Diana—in different countries. Wherever she was worshipped, bees had a semi-sacred status. The Greek father of the gods, Zeus, was a son of the earth-goddess Rhea and Chronos, who usually swallowed his children because it had been foretold that one would depose him. A stone was substituted for the infant Zeus, who was hidden in a cave on Mount Dicte in Crete and reared by the daughters of the king, Melisseus; Amaltheia's goats supplying milk and Melissa's bees honey to feed him. The Curetes, the warrior-priests of the Great-Mother, drowned his cries by beating their spears and

Tanging the swarm

1 Scene depicting intruders in Cretan cave attacked by bees, from a Greek vase of about 540BC *British Museum*

shields together in ritual dances. In another version of the story, the bees were attracted by the clashing weapons of the Curetes; and to this day there survives a mistaken belief that the noise of iron and brass objects being 'tanged' together will make a swarm settle. The legend relates that four men discovered the hidden baby while searching for honey in the cave, whereupon their protective armour fell off and the bees attacked them: this is depicted on the ancient vase (Plate 1). The men were then turned into birds. The bees, as a reward for their services, were given the divine gift of reproducing without intercourse between male and female, and the chastity and immaculate origin of bees was still being stated as fact thousands of years later. Local variations of this story offer other explanations of how bees were created. One says that they were bred from hornets and the sun by nymphs and, after rearing Zeus, were allowed to use honey as their own food. Another tells how the god turned a beautiful nymph called Melissa into a bee and gave her the ability to propagate her kind without a male.

Dionysus (the Roman Bacchus) was almost certainly a god of the honey-drink, mead, not wine, originally. He had a strong connection with bees. The clashing cymbals of his wild followers attracted them, and he was one of the people reputed to have taught men the craft of beekeeping. Others credited with this gift to mankind include the Curetes; and Aristaeus, son of Apollo and the nymph Cyrene, who was himself instructed by the Muses. Bees were known as the Birds of the Muses from the belief that if they touched an infant's lips he would grow up with the gift of song or exceptional eloquence. This was said to have happened to Sophocles, Plato, Vergil and others including, in Christian times, the fourth century Bishop of Milan who became St Ambrose and the patron saint of beekeepers.

2 Wooden figure of St Ambrose, fourth-century Bishop of Milan, and patron saint of beekeepers

The annual renewal of fertility in spring was a vital element in Earth-Mother worship; bees, which disappear in bad weather and reappear with the sun and flowers, were an obvious symbol of it. The frenzied dance with which the Maenads (followers of Dionysus) greeted the spring may have derived from the whirling dance of the swarm. It is strange to find the chaste bee always associated with fertility cults. Bees were deemed to be under the protection of two particular gods, whose statues were often put in their apiaries by prudent beekeepers. One was Priapus, the personification of the fructifying power, always represented with a wizened body and an immense phallus; the other was that lustful Nature deity, the goat-legged Pan.

If bees symbolised rebirth, they also had a close connection with death, partly because they inhabited caves which were regarded as semi-sacred entrances to the Underworld. This provided another explanation of the bees' origin: they were the souls of the dead either returning to earth or on their way to the next world, a belief which is also part of the folklore of many North European countries. This idea probably led to the once widespread custom of 'telling the bees' when their owner died. In some places the hives were turned as the coffin left the house; in others, pieces of black crape were pinned to them or portions of the 'funeral bakemeats' put inside or nearby: but the vital thing was to inform the bees that they had a new master and ask them to stay. If this were neglected, the bees would die or abscond. In some districts they were also told of births and marriages and given a share of the festive food.

The main importance of bees in the Christian church was to provide the enormous quantities of wax needed for candles. There is a German legend that bees were sent straight from Paradise for this purpose, and this version of their origin also appears in some ancient Welsh laws, mixed up with practical matters about the ownership and value of swarms.

Though they had no actual religious significance, bees are connected with Christian saints in several stories. St Domnoc started beekeeping in Wales but gave his hives to St David when he returned to his native Ireland. The bees, however, followed him across the sea and were introduced into a country where previously the very soil had been deemed poisonous to them. (A gross libel!) Another Irish saint, St Gobnat, Abbess of Ballyvourney, when asked for help against a marauding local chieftain, turned a colony of bees into armed men who drove him away, afterwards returning to their hive which in the meantime had become a bronze helmet, or some say a bell. (A more mundane version has St Gobnat hurling beehives amongst the invaders and putting them to flight.) One feels these stories were older legends given respectability by being attached to a local Christian saint. In the same way, the patron saint of beekeeping in the Ukraine, St Sossima, who brought bees from Egypt in a hollow reed, seems to have a strong connection with Zosim, the pagan Russian bee-god.

The folktales which associate Christ with the first appearance of bees also have a whiff of heathen magic about them. In one, Christ made bees, and the Devil, trying to outdo him, only succeeded in producing wasps. A Polish story tells how Paul's head was injured by a stone flung at him, and Jesus took the maggots from the putrefying wound and put them in a hollow tree where they grew into bees. According to French legend, the sparkling drops which fell from Jesus' hands as he washed in the River Jordan became bees and were about to fly away when he ordered them to stay together and work for man, in the name of the Trinity.

If the problem of how the race of bees originated gave rise

to fantastic stories, the mystery of how bees actually reproduced inspired others. People had noticed that each colony contained two kinds of lesser bee and one single Big Bee which they called the king—in a man's world it would have been ridiculous to suppose it could be female—but their roles were not understood. It was generally believed that bees were chaste and fetched their young from flowers. Aristotle mentioned this theory four centuries before Christ but it did not satisfy him. He said that some people thought that only drones came from flowers, and that workers were produced by the rulers whom some people called 'mothers'. This explained why only drones were produced if there was no ruler in the hive (which is true, though the reason is different). Aristotle eventually worked out that, as all three kinds of bee were present when there was a ruler and only drones when there was not, rulers produced other rulers and workers; workers produced drones; and drones did not breed. It was 2,000 years before anyone improved on this hypothesis. The tentative suggestion that the king bee might be a queen and mother was not taken seriously; and until this was accepted little progress was made towards finding out how bees reproduced.

Some time later the idea arose that bees could be spontaneously generated in the carcase of an ox or calf, and this persisted almost into modern times. The notion originated in Egypt and was spread by Vergil and subsequent authors. Columella, a down-to-earth Spaniard living in first century Rome, repeated it with the comment that mortality amongst bees was never so serious as to make it necessary. It would certainly have been expensive.

The method of producing the Bugonia or oxen-born bees was as cruel as it was fanciful. An old translation of Vergil's *Georgics* tells how one must take

> a two-year-old bull-calf, whose crooked hornes bee just beginning to bud; the beast his nose-holes and breathing are stopped in spite of his much kicking; and after he hath been thumped to deathe . . . he is left in the place afore-prepared, and under his sides are put bitts of boughes, and thyme, and fresh-plucked rosemarie . . . In time the warm humour beginneth to ferment inside the soft bones of the carcase; and wonderful to tell, there appear creatures, footless at first, but which soon getting unto themselves winges, mingle together and buzz about, joying more and more in their airy life. At last, burst they forth, thick as raindroppes from a summer cloude . . .

One wonders how many of the writers who glibly repeated and elaborated this recipe over the centuries ever tried to test it. In England in the seventeenth century a Mr Carew of St

3 The drone-fly, *Eristalis tenax*, which resembles a bee

Anthony in Cornwall claimed to have done so successfully, using Vergil's directions. An account of the experiment appeared in a book called *The Reformed Commonwealth of Bees* by Samuel Hartlib, published in 1655. In 1704 Nathaniel Bailey's *Dictionarium Rusticum* mentioned generation of bees from a carcase. The Reverend W. C. Cotton repeated Hartlib's account in *My Bee Book* in 1842: as a child he had wanted to perform the miracle, but he did not believe in it in adult life!

Attempts have been made to explain this peculiar story, one suggestion being that the insect produced was *Eristalis tenax*, the drone-fly. Anyone who has watched these flies crowding the Michaelmas daisies and golden rod in summer will admit they look very like bees (Plate 3). However drone-flies breed in shallow water full of decaying vegetable matter, not in rotten flesh, and the flies which breed in carrion do not resemble bees. In any case such 'bees' would not cluster as Vergil says they did; nor could they be hived as that notable Cornish husbandman (and liar) Mr Carew said they were.

One thing which might have some connection with the story is the appearance of the beast's 'honeycomb stomach'—could its resemblance to real honeycomb have started the tale? Pliny said an ox's stomach smeared with dung would do to generate bees in. This sounds something like the wicker hive coated with clay or dung then in use, and if such a receptacle were dried in the sun it might be taken over by a swarm. The Bible story of Samson and the bees and honey in the lion's carcase becomes just credible if one visualises a carcase hollowed out by carrion-eaters and sun-dried to a leatherlike state.

Gradually, through the centuries, the mysteries yielded to patient investigation: the queen was proved to be mother of every bee in her colony; the role of the drone was clarified; the phenomenon of the drone-laying worker which puzzled Aristotle was explained. We do not yet know everything about the bee, but we have learned enough to discredit all the old myths and superstitions. Nevertheless some persist. We know bees cannot hear sounds in the way we do, only as vibrations travelling along a surface with which they are in contact, but the notion that beating metal objects together will attract a swarm and make it settle has barely died out; and when a friend of mine, answering questions on a radio 'phone-in' programme in 1973, said that bees could not hear so it was useless to 'tell' them about family deaths or anything

4 Leaf-cutter bee gathering pollen from a dog-rose

Leaf-cutter bee's nest and cut rose leaf

else, two listeners indignantly defended the ancient custom.

It must be remembered that what we now regard as picturesque myths were in their day accepted as literal fact. Only simple peasants could believe that bees were the tears of Christ: educated people knew that famous men wrote about bees before Christ was born, but most of them readily accepted the equally mistaken theories of these ancient authorities. Progress in bee knowledge was made by the few who questioned the established beliefs.

The truth is of course that bees evolved over millions of years, like other insects. Honeybees are the most advanced of the many different sorts of bees which exist. Most primitive are the various kinds of solitary bee which live alone and construct cells in which to lay eggs and leave food supplies for their young, which they never see. There are no workers; all females are capable of motherhood. The Leaf-cutter bee *(Megachile sp)* is a typical solitary. She cuts neat round or oval pieces from rose leaves (not petals) and with them fashions cylindrical cells in a tubular hole in rotten wood. A gardener is apt to wonder what caterpillar has such symmetrical feeding habits, unless he happens to notice a small bee flying away with a green disc held in her feet. When the cell is complete, it is provisioned with a mixture of nectar and pollen, an egg is laid in it, then the mother bee puts on the lid and builds another cell on top. She will be dead long before any of her young emerge.

Further up the evolutionary scale are the stout furry bumblebees *(Bombus sp)* familiar to everyone. The young are reared in a communal nest so they are called social bees, but every autumn the whole colony dies except for the fertile young queens who sleep till spring in some hole or crevice. When the weather is warm enough they emerge and each starts a new nest, often in an abandoned mousehole. The queen collects some food, lays a cluster of eggs, and incubates them rather in the manner of a hen. When they hatch, she herself gathers food for the larvae, but when these have developed into adult workers, they take over the foraging and the queen remains in the nest and lays eggs. Later eggs are deposited in wax egg cells, not a brood clump. At its strongest, the colony may number about two hundred individuals. Bumblebees store only enough honey to carry them over a few days of bad weather.

Honeybees have developed still further. Their colony is perennial and numbered in tens of thousands. By storing enough honey to feed themselves and clustering together for warmth, they are able to survive the winter as a community. As there are always workers available, the queen has lost her ability to make wax and to carry pollen and nectar, and has become a highly specialised egg-producer. She cannot live alone because the workers do everything for her, even feeding and cleaning her.

Almost certainly, honeybees first evolved somewhere in southern Asia. There are only four species of honeybees, and three of them, the Giant honeybee *(Apis dorsata)*, the Little honeybee *(Apis florea)*, and the Eastern honeybee *(Apis cerana indica)* are found wild in that part of the world. The Western honeybee *(Apis mellifera)*, the European hive-bee, is very like the Eastern honeybee though larger, and both probably derived from a common ancestor. This bee, as it spread westward and northward through adjoining territories, would gradually develop new traits and adapt its behaviour to cope with the different climates and vegetation it encountered, so that over thousands of years it became a recognisably separate species. The Eastern honeybee, remaining in the original environment, would be closer to the common ancestor. At one time Europe, Asia and Africa were one vast continent, but as the terrain altered, bees which were cut off by new seas or mountain ranges would tend to interbreed and develop regional characteristics, giving us the distinct races, ie golden Italians, European black, greyish Caucasian, which we know today. The Giant and Little honeybees of Asia build single combs in the open air, never in caves or hollow trees as the Eastern and Western species do. It is this preference which has made it possible to keep the latter in hives.

There are no indigenous honeybees in the American continent, though the Maya did to some extent domesticate the stingless bees *(Meliponinae)* found in the sub-tropical areas. The history of beekeeping in America began when European colonists took bees across the Atlantic in sailing ships, probably early in the seventeenth century. The eighteenth-century statesman Thomas Jefferson mentioned in his *Notes on Virginia* that the Indians believed bees came from Europe; they called them 'white man's flies' because they were always seen a little in advance of the settlers as they penetrated further into the country, and the Indians and buffalo retreated. To begin with, they were rare and costly: in Massachusetts in 1660 a stock of bees was valued at £5, a considerable sum at that time.

The honeybee is not indigenous to Australia and New Zealand either, although Aborigines collect honey from the nests of wild stingless bees. Black bees from Europe reputedly reached Australia in 1822 on the convict ship *Isabella.* White men and sheep had arrived thirty-four years earlier. £4 was paid for a hive of bees in Jervis Bay, New South Wales, by a settler who then hired an Aborigine to carry it forty miles to his home. This was in 1840. The first Italian bees to arrive in Australia were sent out by T. W. Woodbury of Exeter, England, in 1862: the voyage took 79 days.

Missionaries are believed to have taken the first bees from Australia to New Zealand in 1839. In 1844, the Reverend William Cotton went out from England. He had written *My Bee Book* two years earlier while at Oxford, and in it he explained exactly how he proposed to carry his beehives to the Antipodes aboard ship. They were to be packed inside ventilated hogsheads (large barrels) resting on racks above blocks of ice, with cinders filling the rest of the space in the upper

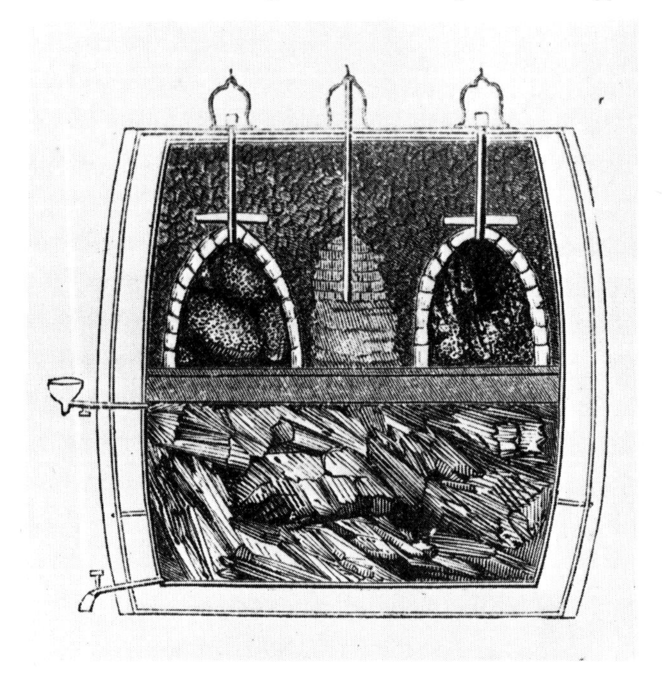

5 Apparatus devised by W. C. Cotton for transporting bees to New Zealand in 1842

half (Plate 5). He tried this, but the superstitious sailors thought the bad weather they encountered was caused by the bees, which they threw overboard, to Cotton's dismay. Once there, Cotton soon produced a *Manual for New Zealand Beekeepers*, for the craft was already gathering momentum. Within a short time it was to develop into a flourishing industry, as it did in America and Australia where climate and flora were equally favourable.

In many ways the late arrival of honeybees in these countries was a blessing. People who began keeping bees in America in the seventeenth and eighteenth centuries did so at a time of great advances in the scientific study of bees, and when real attempts were being made to improve on the antiquated methods and equipment which had scarcely changed since before the time of Christ. They were probably less hampered by centuries of tradition and superstition and so more receptive to new ideas. Certainly in the nineteenth century a flood of new inventions poured out of America and opened the way to modern methods of apiculture.

This beekeeping revolution almost coincided with the arrival of bees in Australia and New Zealand, so that people there could from the start equip themselves with the most up-to-date hives and adopt the most advanced methods of apiary management. In the Old World, established habits and customs died hard, and even today it is possible to find people in Britain keeping bees in much the same way as their ancestors did in the time of Alfred the Great.

2 The Queen Bee

Honeybees: queen, drone and worker

There are three distinct kinds of bee in every hive or honeybee colony. The workers are the most numerous, often 50,000 of them; the drones may number 1,000 but they are only present during a few months of the year. There is just one queen, under normal circumstances. She is larger than a worker; not so broad and heavy as a drone but much longer. Because of the length of her abdomen, her wings look short, though in fact they are not.

As mentioned in Chapter 1, the ancients thought of the queen as a king, a superior and splendid being who ordered the lives of his people, and whose word was law. It was obvious to them that there must be some directing intelligence behind such a marvellous organisation, or else how would each bee know what to do and when to do it; how could they all be persuaded to combine their efforts for the common good? Even when the Big Bee was recognised as female, she was regarded as an Amazonian queen, the absolute monarch of her populous world. Nothing could be further from the truth. The queen is indispensable but she has no power to direct.

Mother-bee, the Germans call her, and this is a better title than queen for she is indeed the mother of her colony. Once she has mated she will lay eggs, probably 2,000 a day during the height of the summer, because the bees must breed at this rate to maintain the strength of the colony. The queen is a specialised machine for egg-laying and, so that she can devote all her energies to this end, she is spared all other duties. The queen bumblebee must be able to construct cells for her young, brood the eggs and carry food for the larvae, because she alone survives the winter and must herself raise the first members of the new colony. Once, queen honeybees were able to do the same, but now that the species has developed a system whereby the community itself survives from year to year, there are always workers available to do the chores. The queen's physical structure has become so modified that she is now unable to do these tasks, and indeed she could not even exist on her own.

A virgin queen may be little bigger than a worker, and

6 Queen with workers, showing difference in size

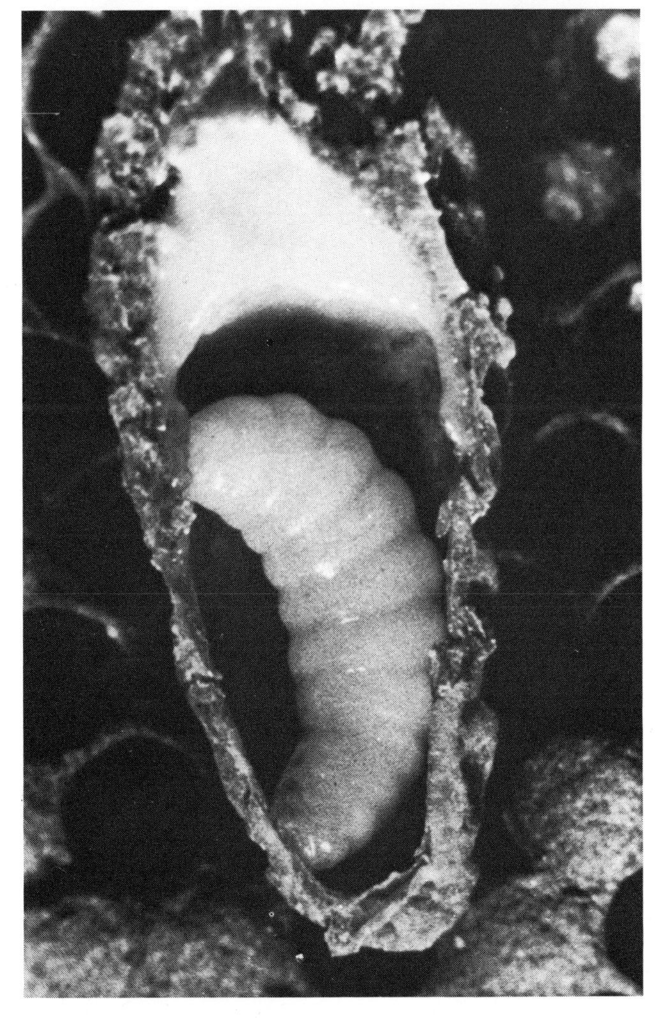
7 Sealed queen cell opened to show larva and royal jelly

both are female, but in all other respects they are very different. The queen lacks the glands which in the worker produce wax, and brood-food for feeding the larvae; her tongue is too short to gather nectar from flowers; there are no pollen-baskets on her long smooth legs; her large compound eyes are each composed of 3-4,000 generally hexagonal facets but the worker's contain many more; her sting is curved instead of straight and is never used except against other queens. The queen's face resembles that of the worker but her brain is smaller. What she does have, unlike the worker, is fully developed ovaries. As a virgin, the queen may help herself to honey from the comb but after she has mated the workers will always feed her. They will surround her, groom her and remove her faeces; they will defend her, with their lives if necessary, and if the colony is starving she will receive the last mouthful of food, for its future existence depends on her. If she is taken from the colony, the workers will follow her, find her and remain with her. They know very quickly if she is missing: within half an hour or less every bee in the hive is aware of the fact. The queen has no maternal instincts. Her young are cared for by their older sisters and she shows no interest in them. All that is required of her is that she shall lay eggs and ensure the colony's continuance: should she fail, she will be ruthlessly replaced.

The egg which becomes a queen is exactly the same as an egg which becomes a worker. The treatment given to the larva which hatches from it alters its development. An egg intended to be a new queen may be laid in a special 'queen cup'. These are extra-large cells shaped like an acorn cup which the workers build at the edges of combs. The tiny, white egg is attached by one end to the base of the cup, and on the third day a minute larva hatches from it. The workers feed it lavishly with special milk-white brood-food, also called royal jelly, and it grows rapidly. All honeybee larvae have this food for the first three days after hatching, but on the fourth day drone and worker larvae are switched to a diet of bee-bread, which is pollen mixed with nectar or diluted honey, while the embryo queens continue to receive royal jelly in huge quantities so that they float on it. The larva is at first curled round, but it soon fills the bottom of the cell and must straighten out, and the walls of the queen cell are extended to accommodate its rapid development. The larva is held on its bed of royal jelly by surface tension, for the cell hangs down from the comb with the mouth at the bottom. It is thought that the extra oxygen available to the queen larva because of her large cell and its free-hanging position—and also because it is made throughout of a porous wax and pollen mixture—may have as much to do with its quick development as the special diet has. On the ninth day after the egg was laid, the queen cell is sealed over by the worker bees. At this stage it looks rather like a large rugged peanut shell.

It sometimes happens that an accident, perhaps the result of a beekeeper's clumsiness, suddenly causes the death or disablement of the queen when there is no egg ready in a queen cup. This is not a disaster because any worker egg, or even a worker larva which is less than three days old, can be reared as a queen. The workers do not take the selected egg from the cell in which it was laid, but tear down the walls of the surrounding cells and begin building a queen cell in their place. This 'emergency queen cell' will not therefore be at the side or bottom of the comb like a purpose-built queen cup, but may occur anywhere on the face of it where there happens to be a suitable egg or small larva.

Inside the queen cell, which has now been sealed by the workers, the larva stretches out and rests, probably absorbing some more of the royal jelly through its skin, and then spins a cocoon around itself and turns into a pupa. The blind and legless grub develops limbs, eyes, wings, antennae, complicated internal organs and a hard shell. The creature remains ivory-white for a time, but as the pupa ages the eyes turn pink, then purple, and as the chitinous covering of the body hardens, pigmentation spreads over the whole insect. On the sixteenth day the young queen begins to nibble away the capping of her cell. The worker bees, knowing to the hour when she is ready to emerge, will have thinned away the wax at the tip of the cell. The queen cuts three-quarters of the way round, the cap falls down like a hinged lid, she unfolds her cramped legs and crawls out onto the comb. Very soon the workers will have torn down and carried away all traces of her cell, for each royal cradle is used only once.

The provision of a new queen is of desperate importance to the colony so anything from two to twelve queen cells may be constructed, although eventually only one queen will be allowed to remain. If the hive is congested and the bees have raised a new queen as preparation for sending out a swarm, the old (ie laying) queen will probably leave with the swarm as soon as the first queen cell is capped, weather permitting. For a few days beforehand, the workers will have restricted the queen's food to cut down her egg-production and enable her to fly. A queen in full lay is unwieldy and if shaken off the comb will flop to the ground helplessly. She may be able to crawl back into her hive but she will not fly back. It is a mistaken idea that the queen leads out the swarm. Usually there are many workers on the wing before the queen emerges; and sometimes she does not come out at all, so the swarming bees are forced to return, and try again another day. If the bees are replacing a queen whose fertility is deteriorating because of age, she might be allowed to remain and continue egg-laying alongside her daughter for a while.

When the prime, or first, swarm has gone, or the colony has lost its queen in an accident, the future prosperity—indeed, the survival—of the colony is locked up in one of those sealed royal cells. The workers wait anxiously for their young queen to emerge. There is a peculiar atmosphere of tension in the hive. The bees do not work so diligently. They lack their normal air of purpose and are inclined to be touchy about human interference. When the young queen pierces her cell and asks for food with protruding tongue, a worker will feed her; but when she at last creeps out and her damp fur dries in the warmth of the hive she may have to feed herself from the honeycomb, for the workers do not make a fuss of her as might be expected.

But what of the other princesses, still enclosed in their cells? The first virgin to emerge may at once seek them out and destroy them. Virgin queens make a shrill piping noise which is answered by those still immured, and this guides the

8 A swarm queen cell; on the right, the domed cap of a drone cell

first-born to the location of her rivals. She tears open their cells and some say she stings the hapless occupants to death, though this is questionable. More probably the workers complete the destruction she has begun. Virgins who have hatched more or less simultaneously will fight to the death when they meet on the comb, grappling savagely with each other, both trying to administer the *coup de grace* with their stings. It is not unknown for both to die of their injuries. In some cases workers have been observed to hide a sealed queen cell by clustering over it and even to delay the emergence of its occupant by repairing the cap as she bites through it, presumably by way of having a reserve.

If the colony is still large even after the departure of the prime swarm, the first virgin to hatch may leave with a second smaller swarm known as a cast (or 'bull' swarm in earlier times). A whole series of swarms may leave at intervals of a few days, the first and largest with the old mated queen, and the others with successive virgins as they emerge, until only one remains to head the depleted parent stock. Occasionally a beekeeper finds a swarm containing two or even three virgins. This might split up, so that each queen heads a new colony; or once the swarm has occupied the permanent site, the virgins may fight until only one is left. The numerical strength of the parent colony and the prevailing weather conditions will decide whether surplus virgins will be destroyed or several swarms sent out, though some strains of bees have an inborn tendency to excessive swarming. These are to be avoided by the beekeeper as only strong stocks can store a large quantity of honey. There are bees which supersede their aged or failing queens and seldom swarm. These are ideal for honey-getting, but the characteristic is apt to disappear in succeeding generations because the new queen may choose her mate from a strain of inveterate swarmers. Unless he lives on an island or in a valley remote from any other apiary, a beekeeper cannot do much (except by artificial insemination) to ensure that his queens mate with what he would consider desirable drones. Their meeting takes place high in the air and some distance from the hives, and he will not see it happen.

The bees are aware of her presence among them, although they take surprisingly little notice of a virgin queen. The future is not yet assured; first she must be successfully mated. The nuptial flight usually occurs between five and ten days after she leaves the cell. In the meantime she wanders about the brood area of the hive, and in the warmest part of the day takes short reconnaisance flights to strengthen her wings and to learn the position of the hive in relation to nearby landmarks so that she will be able to find her way home. Should she enter the wrong hive on her return, she would certainly be killed by its occupants.

The mating flight will take place after noon on a fine day when plenty of drones will be out. As the virgin queen approaches the area where they congregate, they will fly towards her and give chase until one catches and couples with her. His genital organs are torn away and remain in the body of the queen while he falls to the ground dead. The queen may mate with as many as five drones before she returns to her colony, about half an hour after she left it, with the organs of the last drone (the 'mating sign') still attached to her. There is inside her a sac called the spermatheca, which is now stored with millions of spermatozoa received from the drones, enough to last for the rest of her life, which may be five or even six years. She will not fly again unless she leaves the colony with a swarm.

After mating, a change comes over the queen. She is no longer the skittish virgin, apt to fly off the comb if it is disturbed, but a stately matron. Her abdomen is greatly enlarged for now she may lay up to her own body-weight of eggs daily. As she proceeds majestically across the comb the house bees turn to face her, touch her with their antennae, lick her and feed her. The change in the queen's demeanour is matched by the change in the workers' attitude to her. She will never need to feed herself again as she will receive royal jelly from the workers whenever she wants it. She will never again be without a solicitous escort. It is not true—as the older writers claimed—that she has a selected bodyguard or bevy of courtiers always about her; the bees around her are constantly changing. One of the things which makes her easy to distinguish amongst the teeming multitude of her 'subjects' is the sometimes quite perfect oval frame of bees which surrounds her; another is the polished appearance of her thorax which is rubbed hairless by the constant licking of these attendants. These are always the younger bees: the older foraging bees have little apparent interest in her. It may be that there is something rather more sinister about the queen's respectful entourage than at first appears. I have sometimes felt that these courtiers are more in the nature of keepers than sycophants as they hem her in on all sides, penning her on that part of the comb where empty cells must be filled, edging her towards the queen cup in which they will raise the daughter destined to depose her.

Clearly, the queen is indispensable to the colony. It is equally clear that she does not control it nor give the orders, so who does? In a sense it is the workers, though they do not 'decide' anything in a conscious way: what happens is an automatic response to certain circumstances. Weather conditions and the amount of forage available affect bee behaviour. When there is prolonged bad weather and a danger of starvation, the workers react by cutting down the queen's food,

9 An exceptionally large (clipped) queen in full lay

10 A queen with her 'court' of young bees

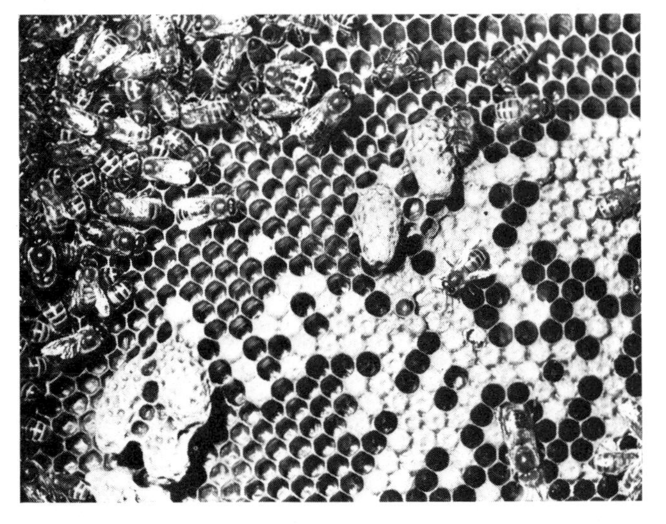

11 Emergency queen cells on comb with capped worker brood

which restricts her egg-laying: they may even drag larvae —especially drone larvae—from their cells and throw them outside the hive to die. They respond to the threat of famine by controlling their population, so that the colony can survive.

The loss of the queen triggers off the impulse to build queen cells, and congestion in the hive causes preparations to be made for swarming. No one issues orders but the key to the bees' behaviour can be found in a glandular product of the queen, known as 'queen substance', so in one sense she is after all responsible for their actions. This acid substance is secreted by glands in the head of the queen bee, spreads over her whole body, and is taken from her by the workers who constantly lick and groom her. As the workers pass food to each other, the queen substance is dispersed throughout the population of the hive.

In effect, queen substance is a sort of tranquillising drug: so long as the bees are receiving sufficient of it, they remain contented and hard-working. It has other properties, such as inhibiting the development of the workers' ovaries. If the supply is inadequate—perhaps because the queen is old or diseased—the workers become discontented and start taking steps to replace her. They build 'supersedure' queen cells. Alternatively, the queen may be producing enough for the colony but its congested state (perhaps due to insufficient hive space, or to bad weather keeping all the bees inside) may be making proper distribution difficult. If a high proportion of the bees are not having a share of queen substance sufficient to keep them happy, queen cells will be built. A complete cessation of supplies, as a result of the queen's death, will trigger off the emergency procedure for replacing her; but if this fails, prolonged lack of the inhibiting drug will cause the ovaries of the worker bees to start developing, and result in the phenomenon of laying workers.

We have seen how queen and worker bees, which are all female, can be reared from identical eggs, but there are also male bees. The eggs from which these come are different in only one respect: they have not been fertilized by spermatozoa from the queen's spermatheca. In some way, which is not fully understood, the queen is able to control which sort of egg she lays, so that every one deposited in a worker cell is fertilised on its passage from the ovary and becomes a worker (unless it is subsequently reared as an emergency queen), and every one laid in a larger drone cell has escaped contact with the sperm and develops into a male bee. Because in fact the drone has no father, a queen who is prevented from mating by bad weather at the crucial time, or because she has some wing deformity, will eventually begin to lay eggs, all of which will hatch into drones. Very old queens, whose supply of sperm has been exhausted, will also degenerate into drone-breeders.

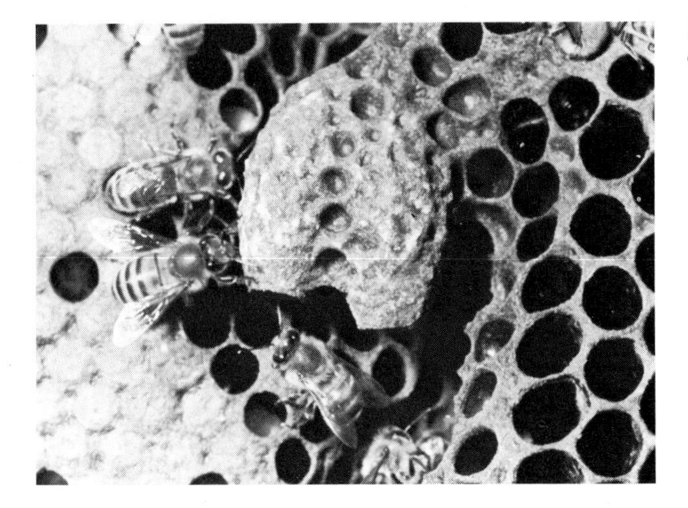

12 Queen cells just before capping

Similarly, because they have not been and cannot be mated, workers in a queenless hive who begin to lay eggs also produce drones. Both drone-breeding queens and laying workers are catastrophic because the bees cannot rear a new queen without having at least one female egg, or a female larva less than three days old. The beekeeper can give them a new queen, or a sealed queen cell from another hive, or some comb containing worker eggs or tiny larvae, and all will be well; but in their natural state they are doomed. In a very few cases, a queen has been known to appear in a colony which has been hopelessly queenless for some time, so it seems that very occasionally a worker may lay a fertile egg, or a female bee hatch from an unfertilised egg, but it is sufficiently rare to be near-miraculous.

So although the queen bee is not a monarch as human beings understand it, she is both more and less. She is the most limited and least versatile individual in the community, but she is also the cohesive force: without her, the orderly functioning of the colony falls into chaos and disintegration. The workers manipulate her but cannot do without her. No one watching her going serenely about her duties among the teeming thousands of her children can be entirely unmoved by her special quality. She may be as much slave as queen, but she has a dignity, almost a glamour, about her. Perhaps the feeling she inspires is the primitive one which found expression amongst ancient peoples in the worship of the Earth-Mother: intuitive respect for the indispensable source of life.

3 The Worker Bees

The worker bee is a miracle of organisation. Her tiny body, scarcely half an inch long, consists of complicated equipment which enables her to perform an astonishing variety of tasks. Everything necessary for the proper functioning of the bee colony, building the combs which form its home, rearing its young, and processing the honey to ensure its survival, as well as defence, cleanliness, ventilation and the collection of raw materials, is the responsibility of the worker. She can do everything but reproduce her kind. To carry out her multifarious duties, each worker is physically equipped with special glands, organs and appendages, so that every part of her is a precision tool adapted to a particular task.

As we have seen, the egg from which the worker comes is identical with that from which a queen is reared. The kind of bee produced is determined by differences in diet and in the amount of space and oxygen available to the larva. Though the queen larva grows much faster, the larval state in all castes is nine days; but after the cell is sealed a queen completes her development in a week, while the worker requires twelve days to reach maturity. This may be due as much to her comparatively complicated structure as to the less favourable conditions in which she is raised. On the twenty-first day after the egg was laid, the young adult worker bites through the capping of her cell and creeps out onto the comb; she grooms herself, and solicits her first meal from a passing sister. At this stage she is pale in colour and fluffy. The ancient bee-masters thought that these were the bees which had grown grey and hairy with age, but in fact the baby down is worn off in the course of their ceaseless labours and old bees have a dark and shiny appearance.

For a while, young bees are what is called house bees. Their work is the care of the queen and larvae and they do not fly outside the hive. At first they are occupied with cleaning out the cells from which they and their sisters have just issued. There is a residue of cocoons and cast larval skins and débris which must be removed, or tamped down, to leave the cell smooth and ready to receive another egg when next the queen passes that way. Worker cells are used over and over again to raise successive generations, and in time they decrease in size from the inevitable build up of compacted débris. The young workers eat quantities of pollen and by the fifth or sixth day of adult life the hypopharyngeal glands in their heads are functional, producing the protein-rich brood-food or royal jelly which is fed to all newly hatched larvae. This then becomes the young bees' work until the supply dries up, usually when they are about ten or twelve days old.

We have seen that the queen honeybee has no interest in her offspring beyond the laying of eggs: the maternal instinct their mother lacks is found in the older sisters of the helpless larvae. This extends beyond members of the same family, for workers given a comb of brood from another hive will care for it as readily as if it were their own. From the third day,

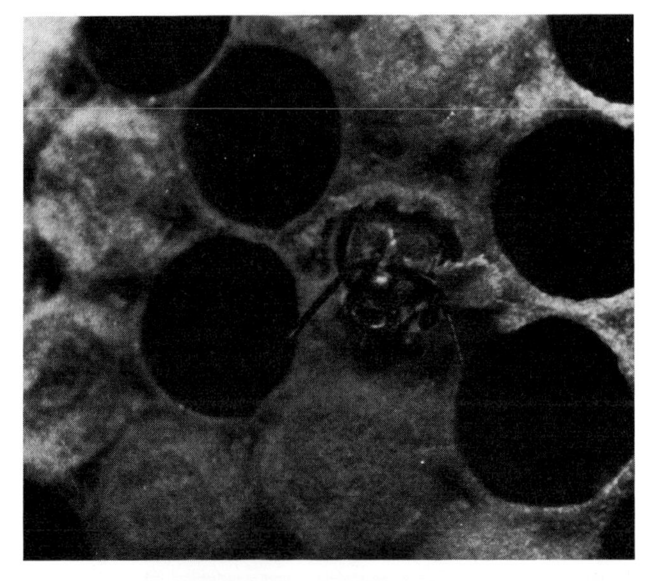

13 Worker bee emerging from her cell

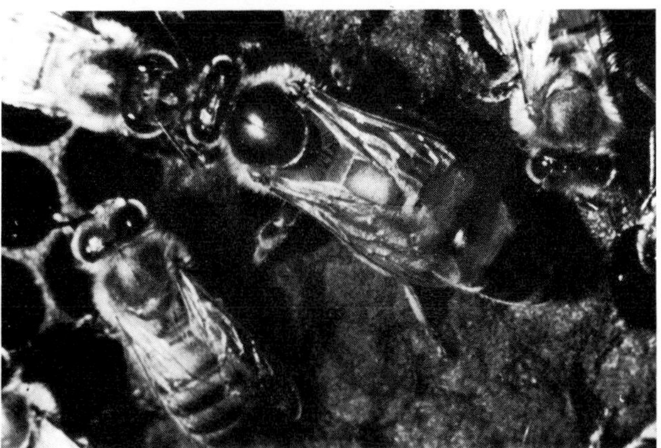

14 A worker bee feeding a (clipped) queen

when the minute larva issues from the egg, the workers bring it food at frequent intervals, probably several thousand times during the six days before the cell is sealed. Worker larvae are never swamped with food as embryo queens are: there is not in any case sufficient room in their smaller cells. They are fed progressively rather than mass-provisioned.

Bee-bread, which worker and drone larvae are given from the fourth day of their larval state, is mainly pollen. This is the protein vital for growth and when brood-rearing is in progress huge quantities are brought in and stored in cells conveniently surrounding those in which the queen is laying. On natural combs (which are curved at the bottom) the brood cells occupy a roughly circular area, though this is

15 Worker larvae, four and five days after egg-laying

distorted to a half-moon shape where combs are built within the rectangular frames in a hive. The pollen stores are arranged in a semi-circle above the brood, and the cells in the upper corners will be filled with honey. A colony may have several combs in which the queen is laying, the central ones almost entirely filled with brood, and the outer ones containing an increasingly larger proportion of stores, so that the brood nest is more or less spherical. The outermost combs of all will be solidly stored with honey. As autumn approaches, the queen lays fewer eggs and the brood area contracts: the worker bees will then fill the vacated combs with nectar from the late summer flowers and the late-flowering ivy. About February the warmest central combs will fill with brood again

16 Newly built natural comb

as the queen recommences laying; and the nest gradually expands over adjacent combs as the colony builds up strength. The beekeeper, watching his hive entrances on sunny March days and seeing workers going in with bulging pollen baskets, is reassured that the queen has survived the winter and is once more in production.

By the ninth day, the glistening white larva fills the base of its cell, which the workers now cover with a flat porous lid of wax and pollen. When comb is built, the cell rims are made thicker so that there is sufficient wax already there to form the cap when it is needed. Inside the cell the larva stretches out lengthwise, spins a cocoon around itself, and changes into a pupa. Limbs, wings, and internal organs develop, colour creeps into the ghostly though perfect insect, and another fluffy young worker emerges to take the place of one who has died, worn out with toil.

As the worker bee's brood-food glands diminish, the wax-secreting glands become active, and she is then capable of building comb. To do this, the wax-makers first gorge themselves on honey, then hang in living festoons with their feet locked together, while tiny flakes gradually form in the eight wax pockets which each worker has on the underside of her abdomen. As the flake appears, the bee takes it from the pocket with her hind legs and passes it to her mouth. Here several flakes are chewed to softness before being fitted into

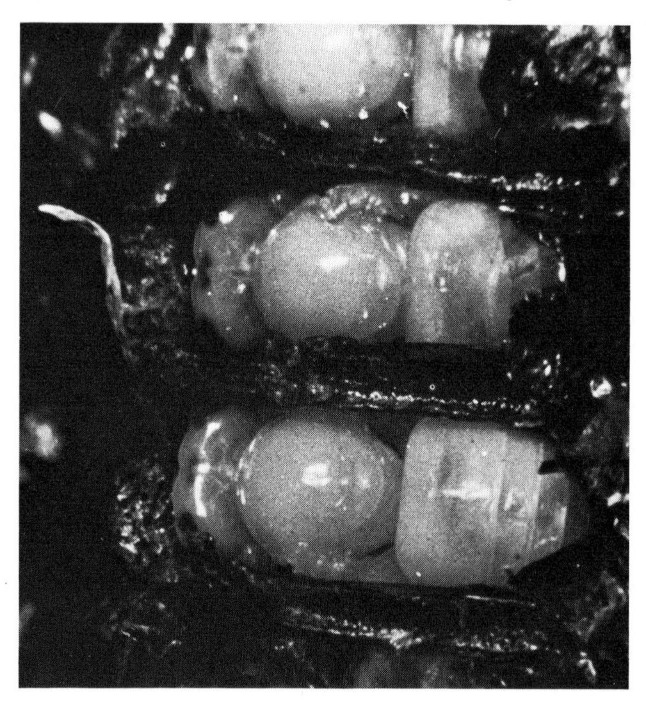

17 Sealed worker cells opened to show pupae

18 Wax scales and partly chewed particles of wax dropped by bees

Underside of worker, showing wax pockets

place. Another lump is added, and another. The speed of construction is surprising. Sometimes wax flakes are formed so rapidly that many are dropped and can be seen lying on the hive floor, looking rather like dandruff (Plate 18).

While some house bees are feeding larvae, and others are repairing combs or building new ones, a large number are occupied in taking the loads of nectar and pollen from incoming foragers and storing it in cells. Nectar is very fluid compared with honey and a great deal of work must be done to evaporate the surplus water and bring it to the proper density. Watery honey quickly ferments so the bees will never cover it with their wax lids until it is perfectly ripe. During the conversion of nectar into honey, important chemical changes take place (see Chapter 6). At the same time, other bees are relieving the propolis-carriers of their loads, of which they cannot rid themselves without help. This sticky resin, fetched from trees in the bees' pollen-baskets, is not stored but is used to fasten things together and also to fill cracks and crevices which the bees cannot get into, both to stop draughts and to prevent pests such as wax-moths (whose larvae eat combs) from laying eggs in them. Another necessary chore is the removal of any débris, including the bodies of bees which have died in the hive, or any interloper which has been killed inside. Occasionally something like a mouse or a mole dies (or is killed) inside a beehive, and presents a problem as the bees cannot move so large a corpse. They have an answer: by coating it with an airtight covering of propolis, they are able to prevent it decaying and polluting their home.

At certain times, many bees will be involved in controlling

the ventilation and temperature of the hive. The brood-nest must be maintained at 31°–35° C (88°–95° F) and the bees' bodies produce the necessary heat. In hot weather there is always a danger that the temperature might rise higher, which would cause the combs to melt and fall. By fanning their wings, strategically placed groups of bees can ensure an adequate circulation of cool air. If there is still a risk of overheating, they can fetch water in their honey-stomachs, regurgitate it onto the combs, and by fanning evaporate it to bring down the temperature.

The guards are probably drawn from the ranks of the foragers, or those bees which are about to begin foraging. A worker will be about twelve days old when she leaves the dark hive for the first time and ventures into the sunlight. Some warm afternoon she will make a little flight close to the hive, and repeat it several times during the next few days so that she becomes familiar with its position in the landscape. Many workers share the same birthday, due to the queen's prolific egg-laying, so a whole batch of young workers will take to the air at once, giving at first glance the impression that a swarm is rising, but soon they retire to the hive again. On her first orientation flight the bee will relieve herself for the first time, for no healthy worker excretes inside the hive: it is possibly the discomfort of her distended bowel which drives her out.

It must not be thought that every bee does every task in strict rotation and at exactly the same age. The division of labour is flexible. Should there be a gap in brood-rearing due to the loss of a queen and the raising of a new one, young bees will join the foraging force much earlier. If there is a shortage of young bees to feed and cover the brood, some foragers will

19 Guards at the hive entrance

revert to household duties. Sudden damage to a comb will draw bees from other occupations to help in the urgent work of repair. Some young bees fly out of the hive long before their twelfth day too, for they can be seen, conspicuous in their grey down, in swarms. Any individual bee may miss out one or more tasks entirely, or do several concurrently. She may clean out any dirty cell or repair any broken one she comes across in the course of feeding larvae, or leave her pollen packing to help drive out an intruder. They seem to obey to the letter a Victorian text which, years ago, hung on my nursery wall: 'Whatsoever thy hand findeth to do, do it with all thy might'.

Most bees are probably three weeks old when they begin foraging, and they will spend what is left of their short lives—probably three more weeks—toiling back and forth from dawn until dusk with loads almost as heavy as themselves. Bees are very strong since they are able to carry another (dead) bee some distance from the hive before jettisoning it. It has always astonished me that the first few bees of the swarm to settle must take a grip firm enough to support the weight of the remainder of the cluster; and a swarm can easily weigh six or eight pounds. To airlift their loads of pollen and nectar, they may have to contend with unfavourable winds which must often damage their fragile wings, and they make continuous trips during the hours of daylight. A worker will fly up to two miles from her hive to a good source of food if the weather is good. There is a carefully computed relationship between the food required to fuel the bee till it reaches the foraging ground and the amount of food it will be able to bring back: bees do not make uneconomic flights. When the temperature falls below 10° C (50° F) honeybees are unable to fly, though they can still walk—at 7.7° C (46° F) their legs

20 Worker collecting nectar from a blackcurrant flower

also become paralysed—so early and late in the year when the weather is liable to change suddenly, they do not fly far from home. The normal speed of a worker bee is around 12 mph and her direction-finding is fairly miraculous. The sun—or at least the polarised light which she is able to distinguish—is her compass; but some mechanism takes note of all her seemingly random movements from flower to flower, for when she has her full load, after visiting perhaps 1,000 flowers, she returns unhesitatingly to her hive. Bees do sometimes lose their way and enter the wrong hive, especially where identical hives are regimented in straight rows without trees or bushes by which individual ones may be distinguished. A *laden* worker from another colony would probably be allowed in, although a strange queen would be killed. Considering the number of bees and their daily trips, mistakes are rare.

While the bee is carrying nectar homeward, glandular secretions in her honey-stomach have already begun the process of converting it into honey, a task which will be continued by the house bees. The heat of the hive also helps to condense and ripen it. Linnaeus named the honeybee *Apis mellifera* (honey-carrier) in 1758, but in 1761 he changed this to *A mellifica* (honey-maker). We have now reverted to the older name, which seems to me wrong. Bees carry nectar: they manufacture honey. Nectar is a raw material, just as fruit and sugar are the raw materials from which jam is made. Much of the manipulation of the nectar, the moving of semi-processed honey from cell to cell to help evaporate the surplus water, is carried out during the warm summer nights, for it is a mistake to suppose that bees return homeward at twilight to sleep. The work of the hive goes on without pause. When the honey has reached its proper consistency it is capped with wax, and in this state will keep for a very long time.

21 Pollen-carrier resting before entering the hive with her load

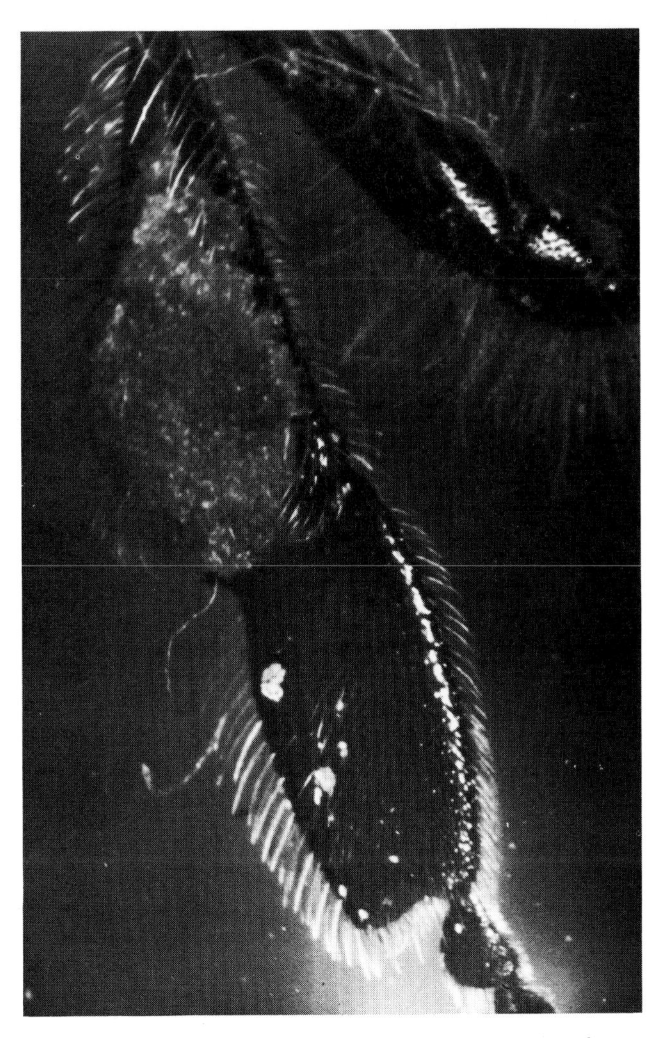

22 Hind leg of worker bee showing pollen basket and load

Pollen is generally easier to collect as it occurs in larger quantities. The bee's body, even her eyes, are covered with hairs to which the pollen adheres as she pushes her way in and out of the flowers. Her legs are equipped with brushes of stiff bristles with which she gathers the pollen together; it is then moistened with a little nectar so that it sticks together, and the resulting paste is packed into the pollen baskets on the hind legs. The basket is formed by the smooth concave *tibia* of the bee and the fringe of stiff hairs which surrounds it. When the forager staggers in with her load she lowers her back legs into a cell, pushes off the coloured balls of pollen with her middle pair, and then returns to work. A house bee breaks up the lump, rearranges it in the bottom of the cell—or another one—and rams it down with her head. Pollen which will not

23 Comb on joint of bee's hind leg which removes pollen from the opposite leg

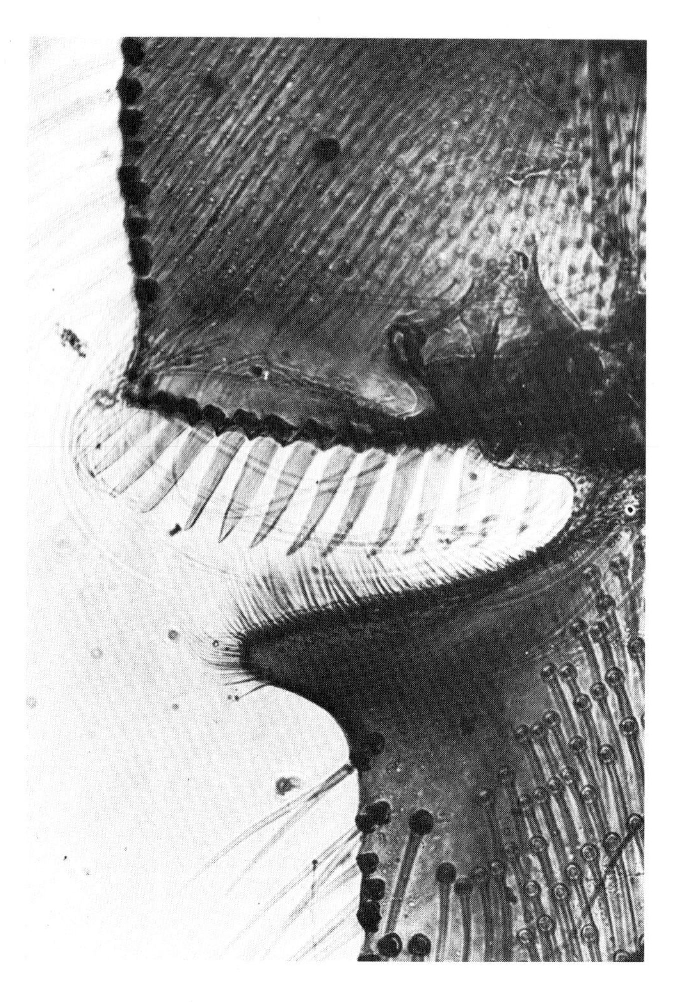

be eaten immediately by larvae, or by young bees producing brood-food, is preserved under a layer of honey. This will be capped with wax if it is intended as winter stores.

No one knows how it is decided which bees will accompany the old queen when she leaves with a swarm, and which will stay behind. Bees of all ages comprise the swarm. Before leaving, the workers gorge themselves with honey, which is why swarming bees are usually good-tempered and disinclined to sting (Plate 24). This food supply ensures that the bees will survive till they find a new home and can start foraging again; and also that they are primed for wax-building in the new site, for wax can only be produced if the bee is carrying a load of sugar surplus to her own physical requirement.

Occasionally scouts go out from the hive to look for a new home before the swarm leaves, but this usually happens after

24 Newly swarmed bees are gorged with honey and good-tempered

25 A swarm of honeybees in a nut tree

Worker bee carrying pollen

the swarm has clustered on some convenient tree or fence nearby. Bees may return with news of several potential nest-holes but unanimity must be reached before the swarm will move. A bee can communicate the whereabouts of the site it has found in the same way as it conveys information about the location of a food supply—by dancing. The bee performs a series of circling movements interspersed with straight runs (which indicate the direction as an angle from the sun's position), and other bees following her movements learn the distance and route to the find. It is possible to see bees treading their regular pattern on the surface of the swarm cluster. What is more extraordinary is how they eventually agree which of the possible sites they will occupy. When this happens, the whole body of bees takes off and flies directly to the chosen place. (The proverbial 'bee-line'?) This may be one hour or three days after the swarm left the hive. The longer the bees remain where they first settled, the more irritable they become, as their reserves of honey are used up. Sometimes a few scouts will return after the swarm has moved off (or been removed by a beekeeper), and will hang about looking for their departed queen and becoming increasingly touchy until they give up and return to their original home. Sometimes the scouts cannot find a suitable cavity for the swarm to occupy and, under pressure of the wax-making urge, the bees begin building comb on the branch where they have clustered (Plate 26). A colony may build several combs in the open, raising brood and storing honey quite successfully in the summer months, but cold winds and rain will destroy it when winter comes.

As soon as some comb is ready, the queen—unless she is a virgin—will lay eggs in the cells; indeed, she will lay eggs when the cell walls are hardly begun and the workers will complete them around the developing larvae. Food must be stored as soon as there are cells available to put it in, for should a spell of bad weather intervene, the colony, having no reserves, must perish.

A beekeeper collects a swarm in a box or straw skep and, having prepared a hive, simply dumps the whole mass of bees on a wide board sloping up to the entrance. In a few seconds the bees begin to run up and disappear inside. The queen strides rapidly over the moving stream of workers and enters the new home. Around the entrance certain bees put down their heads, elevate their tails, and fan their wings so briskly that they become almost invisible. Sharp eyes can distinguish on each fanning bee a small white mark towards the end of the upper surface of the abdomen: this is the Nassenov gland. When open like this, it produces a scent which is dispersed by the wing movements and acts as a signal and summons to the other bees. As the stream of scent flows back over the main

26 Combs built in a holly bush; such a nest would not survive the winter

27 A colony of bees established in the roof of a building

body of the swarm, more and more bees hurry up into the hive and a satisfied hum comes from inside as they take possession. Bees which have found a 'lost' queen will use this method of indicating her whereabouts to the rest of the colony.

There is probably a characteristic hive-scent by which guard bees can distinguish their own foragers from strangers. Two colonies of bees can be united by removing one queen and putting the two hive bodies on top of each other, separated by a sheet of newspaper: by the time the bees have eaten through the paper (and removed it all in minute shreds to the outside world through the hive entrance), all will have acquired the same hive-smell and will function as one community. Any attempt to put strange bees straight into another colony will result in a battle royal and thousands of dead, though two swarms can under certain circumstances be safely amalgamated.

A bee's sense of smell, which is located in the last eight joints of the antennae, helps to guide it to flowers which provide nectar and pollen. Bees are also well endowed with eyes: two large compound ones each composed of 6–7,000 facets, and three small simple eyes at the top of the head called *ocelli* or *stemmata.* The latter probably have no vision, but can detect changes in the intensity of light. Even with their large eyes, bees do not see things as human beings do, that is, as a sharply defined image, but rather as an indistinct mosaic. Their different perception of colour is discussed in Chapter 5.

There is one other feature of the bee which commands great, though wary, human interest: the sting. This is in fact the ovipositor of the female insect which in worker bees has lost its function as an 'egg-placer' and has become instead an effective weapon of defence (not offence). While a virgin queen will fight another on sight, workers do not fight within the colony but keep their stings for use against human and other animals which molest them, or interlopers which force a way into their home, such as wasps, alien bees and mice. Away from their hive, bees sting only if they are trodden on or picked up, that is, when they are hurt or frightened. If a hive has been disturbed, there may be angry bees flying around some hours later, which are inclined to assume evil intentions on the part of anyone passing. Even in the hive, the temper of the bees varies according to conditions. On a warm day when the nectar is coming in fast, the colony is too absorbed in its work to take much notice of the beekeeper's intrusion. A change of weather to cold wind or thunder, the end of a honey flow when a crop passes its flowering period, or a domestic crisis such as the loss of their queen, may upset them and make them irritable. Human beings who are suddenly deprived or threatened react in the same way.

When a bee stings a person, it forfeits its life. The sting is barbed, and once it is thrust into the comparatively thick human skin the bee cannot withdraw it. It twirls about until it tears itself free, leaving behind the sting and parts of its body. The injury is always fatal to the bee: the human being recovers, except in the case of a very few people who are allergic to bee venom. People who blame the bee for being 'vicious' might consider who is really the aggressor, and perhaps appreciate the courage of the tiny creature who has surely a right to defend her home and hard-won treasure. Thousands of her sisters have died in the garnering of their harvest: a few more will die to protect it, since it guarantees the survival of the colony.

As the autumn days shorten and the nights become colder,

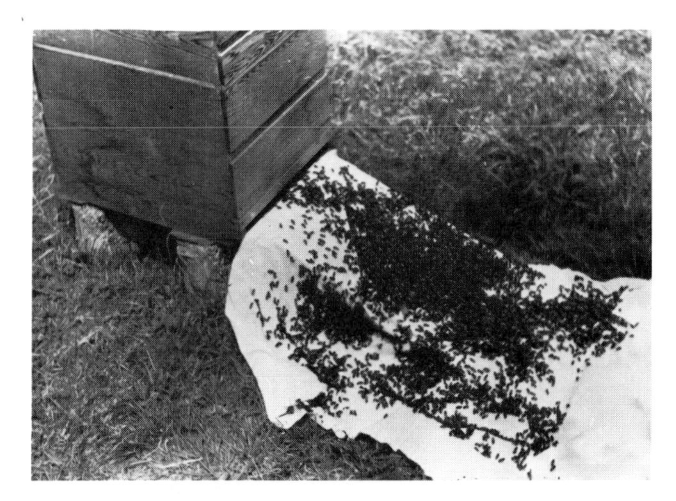

28 A swarm entering a beehive

29 Worker bees fanning with Nassenov gland exposed

many of the older bees die. It is the young workers emerging from their cells in October which will form the winter cluster and live to rear a new generation the following spring. These autumn-born bees may have a life span of six months compared with the six weeks normal in the summer. A spring or summer-born bee works all her short life; she may enjoy just one day's holiday if she joins the delirious swarm-dance and leaves her home for an uncertain future. Even when bad weather confines her to the hive, the work goes on, for there are always larvae to be cared for, stores to be moved and sealed.

As soon as the winter begins, the routine labours of the hive come to an end. The queen has gradually stopped laying so the brood cells are empty and every available corner is stored to capacity with honey and pollen. By eating honey, the bees will generate sufficient heat to keep that part of the hive where they are clustering at a temperature above 20° C (68° F); and in February this will rise to at least 31° C (88° F) on the combs where the queen has recommenced laying eggs. The colder the outside temperature, the closer the bees pack together, for the smaller the surface area of the cluster, the less heat is lost. The cluster may cover two or three combs and the spaces between; as the honey is used up the bees move on, very slowly, still retaining their tight formation. On sunny days, even in midwinter, the bees may leave the hive for a 'cleansing flight' to relieve themselves. So long as they are flying, they are all right, but if they land anywhere they soon become chilled and incapable of taking off again. After a brief gambol in the open air, the bees reform their cluster, probably on a new area

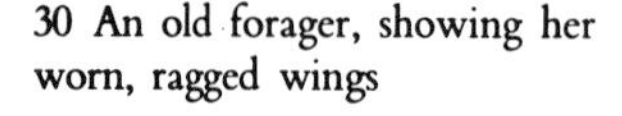

30 An old forager, showing her worn, ragged wings

of comb where stores are plentiful, and it may be several weeks before they can take another flight.

Long before signs of spring become obvious in the outside world, a whisper has reached the inmost parts of the hive. The drowsy workers stir; their queen asks for and receives more food than her bare subsistence; tiny larvae gleam like pearls in the cells at the heart of the cluster. Soon a warmer air will call out the bees to their tasks: some will fetch pollen from the early catkins, some water to dilute the stored honey, so that the increasing host of young can be fed. As the sun strengthens and more and more flowers open, thousands of workers pour out into the bright world to pursue their lives of patient toil.

4 The Drones

This is certain to be a short chapter, for what is there to say about the drone honeybee? He is born; he serves his one indispensable purpose; for a brief while he takes his pleasure in the heat of summer days; and he dies—suddenly and brutally destroyed by his sisters. Ever since he was recognised for what he is, the drone has been the object of derogatory comment—the word itself is used to describe a luxury-loving man living by other people's work. The drone bee is, however, just as essential to the continuance of his species as the other members of his community.

It seems surprising that the sex of the drones was ever in doubt. They are so aggressively masculine in appearance. More heavily built than the queen, though considerably shorter in the body, they have a thick, furry pelt and a fringe of long hairs at the end of the abdomen which gives them a blunt, squarish shape. This is very noticeable amongst the tapering bodies of the female bees. Also conspicuous are their huge eyes which meet over the tops of their heads, so that the little simple eyes are crammed into a bunch almost in the middle of the mask. In contrast to the heart-shaped face of the worker, the drone's head, viewed from the front, is almost round. The huge eyes together containing about 17,000

31 A drone, with workers, showing his heavy build and large eyes

facets—more even than a worker's—give him the sight necessary to pick out a young queen on her nuptial flight; and so that he can catch her, he is equipped with large, strong wings and a top speed of probably about 15 mph.

There are no drones in the colony during the winter. In March the queen will begin laying a few eggs in the larger cells on the outskirts of the brood area, which will have been emptied of honey during the cold months. Drone cells are usually on the edges of the combs, and usually the outermost combs only, so that if the brood area has to be reduced due to adverse conditions, the drone brood will be sacrificed first. It may be that a spell of bad weather makes foraging impossible at a time when the winter stores are almost used up; or the death of many bees from discease, or as a result of crop-spraying with insecticides, may leave the colony short of workers to cover the brood. In either case the bees will concentrate their care on the worker larvae in the centre of the nest and the drone brood around the perimeter will be left to perish. It would clearly be uneconomic to rear a crowd of hungry, non-productive drones in difficult times.

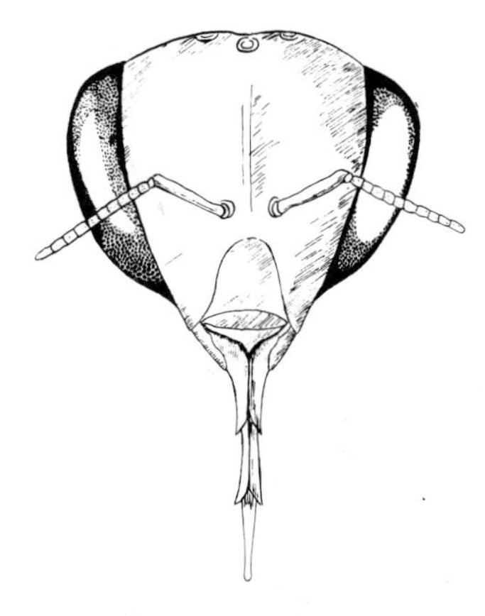

Faces of worker (with tongue extended) and drone

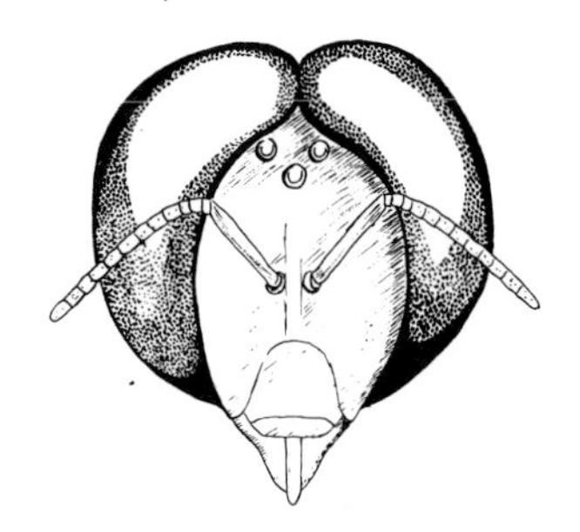

During May and June, the peak months for swarming, it is expedient for the bees to raise and maintain drones because they—or a tiny percentage of them—will be needed to impregnate the virgin queens. If at any time the colony is threatened with starvation, however, the workers will drag drone larvae and pupae from their cells and throw them outside the hive to die. I have seen bees in May prevent the drones re-entering an old-fashioned straw hive after their midday flight; after a chilly night there were still about fifty drones in a bewildered huddle around and above the entrance, which was solidly guarded by determined workers. In a skep containing natural comb, the bees build a higher proportion of drone cells than a modern beekeeper would tolerate and in this case the bees obviously decided they had overdone it. In a frame hive, the amount of drone brood raised in a colony is to some extent controlled by giving only wax foundation stamped with worker-cell bases; though should they think it desirable, the bees will tear down worker cells and build drone comb in its place or vice versa. It has been estimated that one-fifth of the natural comb built by a wild or skep colony would be drone comb, but bees really prefer building drone comb for storing honey in, and much of it would have been built originally for that purpose. In the following year, when the honey had been used, the brood nest might expand to include these combs and the queen would put an egg in every cell, with a consequent huge production of drones.

The drone bee comes from an unfertilised egg and therefore has no father. As the drone inherits genes only from his mother he resembles her in type and colour. Thus a black queen may breed yellow-striped daughters, but all her sons will be black like herself; they will not, of course, pass on any yellow characteristics to their progeny when they mate, since they have no yellow genes. Because of their parthenogenetic origin, drones can be produced by a queen who has not mated at all, or by a laying worker in a queenless hive. Such a colony is bound for destruction as fewer and fewer workers struggle to meet the demands of an ever-increasing drone population. The bees may become so demoralised that they build royal cells around drone eggs—the only sort they have—in the vain hope of rearing a new queen. As drone eggs have been laid in worker cells they are obliged to put domed drone caps on these to accommodate the larger larvae, though in fact the drones which emerge from them are usually undersized. The last workers die; some drones may feed themselves from the store cells for a while but all are doomed and robbers will soon clear out whatever honey is left. A few drones may drift to other hives because for some reason these harmless creatures do not arouse the aggressive instincts of the guards and are permitted to come and go as they please.

So drone eggs are laid, larvae are fed, and in due course the cells are covered with their distinctive domed caps. Twenty-four days after the queen deposited her egg, the fur-coated, swaggering drone emerges. It is still an unsolved question why the drone takes so long to develop, since he is physically a simpler creature than the worker and his sexual organs are already formed in the larval state. The queen, also sexually complete, takes only sixteen days to reach maturity. A drone is not capable of impregnating a virgin queen till about fourteen days after he emerges from his cell. There may be some delicately balanced time-scale intended to ensure that the drones of one colony do not mate with virgins from the same colony, a time-scale which the beekeeper's interference probably upsets. However, the drones do not show any interest in a virgin in their own hive and would be unlikely to be in it when she leaves on her nuptial flight. Once she is in the air, the chances of her mating with a drone from her own colony must be slight.

Once hatched, the drone for a while enjoys a pleasant existence. He asks for, and receives, food from any house bee he passes, or helps himself at will from open storage cells. Drones do no work at all and they excrete in the hive, unlike the workers which clean up after them. It is possible that they serve one other purpose beside the sexual one—probably inadvertently. As they like to congregate in the warmest part of the hive, that is the brood nest, they help to maintain the high temperature necessary to incubate the eggs and keep the larvae warm. This must release many workers for other duties. The drones only fly out of the hive during the hottest part of the day—usually between noon and 4 pm—when there would be little chance of the brood being chilled. These flights are only taken for pleasure and the chance of sexual adventure. Unlike the males of other species of bee, drone honeybees never refresh themselves from flowers but feed only within the hive. Their tongues are not long enough to reach the nectar in the blossom's calyx. This dependence on the hive for their food naturally makes their destruction later in the year very much easier.

There are certain areas where drones assemble, high up in the sky, where they are invisible although their sonorous hum is easily audible far overhead. The Reverend Gilbert White of Selborne in Hampshire wrote about this phenomenon in his journal and the noise of the drone assembly can still be heard today, two hundred years later, in the place he mentioned. Several other such areas have been recorded and in some the drones have been watched through binoculars. Although their existence is established beyond doubt, the significance of drone assemblies is still uncertain. Do they, for instance, occur in all countries where honeybees are found; and what attracts drones to one specific area rather than another? Does the virgin queen fly towards them, and if so, how does she locate them since she is deaf to the droning sound? In any case, it would be a strange reversal of normal behaviour, at least in the insect world, if the female went in search of the male. Are queens and drones attracted to the same area by the same factors? And what could be the purpose of such a gathering since few of the congregated drones will actually succeed in mating a queen? It is a fascinating question.

Mating takes place on the wing, for only in the high spaces of the air are the drones' masculine passions aroused. Many people have seen and described the marriage flight of the queen and have wondered at the excessive number of males which pursue her, most of which can have no chance of intercourse with her. Nature often seems prodigal: consider the number of eggs a fish releases into the sea, only a tiny proportion of which will develop to maturity. As they are food for so many natural enemies, such excess is the only way of ensuring that *some* will survive to perpetuate the species. So I think it is with the drones. Mating bees do not have the skies to themselves: hawks, swifts and many other birds may eat them, and the more drones there are to serve as their prey, the better the chance of survival for the indispensable queen. Dr David Lack, the ornithologist, in his book *Swifts in a Tower*, tells how an observer who dissected a number of swifts to establish whether they ate honeybees, found only drones. It is desirable that the precious queen should mate the first time she flies out to do so, making further dangerous trips unnecessary; and the more speedily she meets the drones, the less time she

32 A piece of comb, showing the difference between drone and worker cells

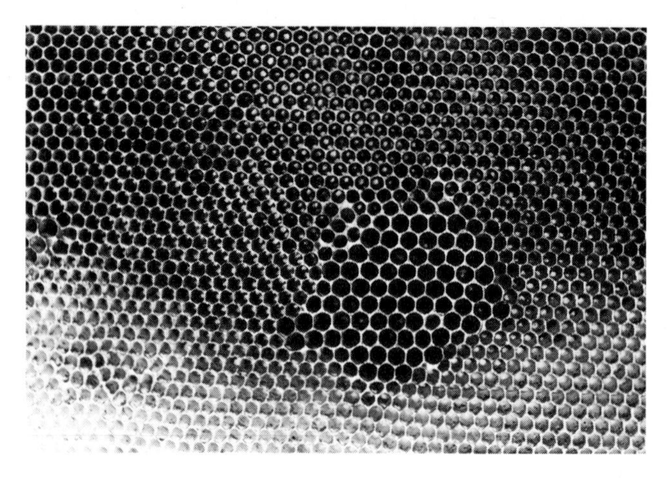

will spend cruising about in vulnerable isolation. With drones available in such quantity, the risk to the queen is much reduced. It was thought for a long time that mating was a once and for always event, but it is now certain that the queen mates with more than one drone, probably about five, before she returns to her colony. The sperm received at this time lasts throughout her life. The drones—the chosen few—die, for their reproductive organs are torn away in the act of fecundation.

With the queen successfully fertilised, the bees have no further need of drones. Thousands of worker larvae must be reared to replace those killed by hard work during the summer's foraging and to make sure that there will be a large number of young bees available to form the winter cluster. Stores must be husbanded for the flowerless months and the greedy males eat too much. Their usefulness at an end, the drones are under sentence of death. In an exceptionally strong colony they may be tolerated longer but their presence in any number in the autumn is a sign that all is not well with the community. Either the bees have a drone-breeding queen, or an accident has deprived them of their sovereign and they are allowing some drones to live a little longer in the faint hope that they may raise a new queen and still get her mated at the eleventh hour.

For the majority, once the colony has settled down after the swarming season, the end is quick and brutal. The workers stop feeding their greedy brothers. The drones try to feed themselves at the open cells but the relentless workers chase them away through the narrow pathways of the hive. They huddle hungrily in corners or fly out into the sunshine. But with empty stomachs there is no pleasure in the warm skies, and when they go home they find their way barred. The implacable workers are hustling their brothers out of the entrance, pushing them off the flight board, and biting their wings to disable them. Weakened with starvation, the harassed drones cannot resist. Their size is no help for thousands are ranged against them, and they are few and unarmed. It is less a massacre than a mass expulsion. Contrary to general belief, the workers seldom if ever sting the drones to death. It is starvation and the damp chill of the nights when they are shut outside the hive, which kills them. In the torpor resulting from cold and hunger, they are easy prey for birds, toads, and other predators. Hedgehogs have been known to establish a regular twilight patrol in August to clear up the discarded drones in front of the hives.

Beekeepers often confess a sneaking sympathy for drones. During their brief playboy existence they do seem to enjoy themselves. The picture of them given by the Reverend Charles Butler in *The Feminine Monarchie* (1609) expresses it perfectly.

> The Drone which is a grosse hive-bee without sting, hath been alwaies reputed for a sluggard, and that worthily: for howsoever he brave it with his round velvet cappe, his side gown, his great paunch, and his lowd voice; yet is he but an idle person living by the sweat of others brows. For he worketh not at al, ether at home or abroad, and yet spendeth as much as two labourers; you shal never finde his maw without a good drop of the purest nectar. In the heat of the day he flieth abroad, aloft, and about, and that with no small noise, as though he would doe some great act: but it is only for his pleasure, and to get him a stomach, and then returneth he presently to his cheere . . .

Their carelessness is in some ways more attractive than the meritorious but alarming devotion to duty displayed by the workers. The drones' big eyes give them an innocent, bewildered appearance which makes their end doubly pathetic. They hang about the hive entrance looking stupid and apologetic as if hoping their heartless sisters will relent and let them in, but the workers never compromise. Nature ordains that if the colony is to survive the winter there must be no useless mouths to feed, so the sentence is ruthlessly carried out.

Even beekeepers regard drones as expendable. Beginners learning to pick up a queen bee between finger and thumb, or to capture one in a matchbox or queencage, first practise with drones, which cannot sting and will not be missed if the operation is bungled. Those who clip their queens' wings so that the prime swarms cannot fly away and be lost, perform this delicate operation on a few drones first, for it is easy to cut off a leg too, and one cannot risk maiming a queen. So the drone serves another purpose Nature never thought of.

The Reverend J. G. Digges, a great Irish beekeeper with the soul of a poet, must have the last word. For centuries clergymen held up the drone's fate to their congregations as a dire warning to the slothful, but Digges was a Christian of another sort. In his book *The Irish Bee Guide,* 1904 (later reissued as *The Practical Bee Guide*) he had this to say of those 'oft-maligned, noisy, buzzing bees':

> Theirs is a life of brief dependence and submission. They gather no stores: nature has not fitted them to do so. The one object of their existence is to fertilise the young queens. To that end they are born, are tolerated in the colony, and are allowed free access to the honey cells. Theirs, also, is the sacrifice of life to duty; and such of them as survive to the close of autumn are driven out of the hive to end, in cold and hunger, a life which, if seemingly idle or useless, was, at least, inoffensive, and full of possibilities whose vastness fills with awe and amazement every thinking mind.

5 Bees and Flowers

Flowers and bees are interdependent, and have been so for a very long time. It was not for the gratification of Man that flowers developed their gorgeous colours and scents, but for the satisfaction of their own need to reproduce. Both are designed to attract pollinating insects, without which the plants cannot set seed or produce fruit, no matter how congenial the climate and soil in which they are growing. It is noticeable that flowers which are wind-pollinated, for instance those of grasses, sedges and some trees, are small, scentless, and inconspicuously coloured in brown or green.

Pollination is the transference of the male germ or pollen from the anthers of a flower to the female parts (stigmas) so that they are fertilised. Once they have made contact with the stigmatic surface, the pollen grains germinate and send down tubes through the style to penetrate the ovules contained in the ovary. Pollen may come from the same flower *(autogamy)*, from another flower on the same plant *(geitonogamy)*, or from another plant altogether *(xenogamy)*. Ideally, cross-pollination involves flowers of two plants or even varieties of plants of the same species. For instance the apple called Cox's Orange Pippin will not fruit at all unless it receives pollen from another variety blooming at the same time, such as Charles Ross, Grenadier or Golden Delicious. The self-fertile variety

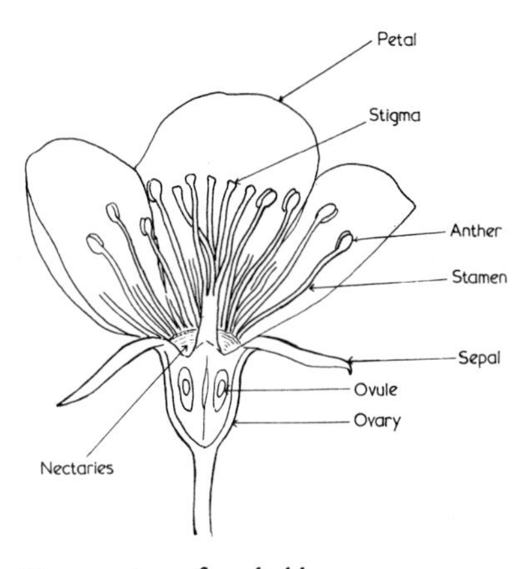

Cross section of apple blossom

33 Hairs all over the bee's body pick up pollen from the anthers of the flower

Bramley's Seedling will bear fruit on its own, but produces a heavier crop if it is cross-pollinated by another variety. Under no circumstances will an apple and a pear tree pollinate each other's blossoms. Both self-pollination and cross-pollination are carried out by insects, of which honeybees are by far the most efficient.

Bees depend entirely on plants for their food and that of their young. Nectar is mainly carbohydrate—the honey made from it contains only 0.2 per cent protein. Pollen is the protein-rich food which ensures the rapid development of the larvae. Only if bees perform the task of pollination can the plants propagate themselves, but only if plants produce nectar and pollen to feed them can bees survive.

Bees and flowers almost certainly appeared at the same time, millions of years ago, and each influenced the development of the other. The earliest flowers had no petals, and may have developed them mainly to protect their pollen from rain. Bees evolved from a wasp-like ancestor, which visited plants originally to catch prey for its carnivorous larvae, but gradually took to feeding them on nectar and pollen. When coloured petals occurred, bees probably visited the more conspicuous flowers for preference, and these, being better pollinated, were swifter to increase, so eventually superseding the others. The bees began developing hairy bodies to collect pollen, also tools on their limbs to gather it up more efficiently, and longer tongues to extract the nectar more easily; and the bees which had these modifications were more successful, bred faster, and became the dominant kind. So structural changes eventually produced flowers ideally suited to supplying bee food; and bees well adapted for pollinating them. In some cases a very close bond exists between a flower and a bee, the classic one being the association between the little solitary bee, *Halictoides novae-angliae* and the pickerel-weed *(Pontederia cordata)*. The flowering of the plant exactly coincides with the emergence of the adult bee, which works it exclusively, although other suitable flowers are available at the same time.

Insects other than bees which play a part in pollinating flowers are flies, butterflies and moths, and some of them visit the same flowers as bees do; but as they are only looking for nectar to drink themselves and will flit from one plant to another indiscriminately, they are not ideal pollinators. Bees have to gather very large quantities of pollen to feed their young, so they will visit many flowers on each trip. They are also far more systematic in working one species of plant at a time, though bumblebees are less reliable in this respect than honeybees. Because a new colony is raised each spring from an overwintered queen, bumblebees are never numerous in the early months of the year, and after a severe winter may be

34 Nectar-gatherer with tongue extended

very scarce. Some varieties of wild bee are active only for a month or two in the year. Honeybees are available in large numbers whenever there are flowers to be pollinated, and their role as pollinators of fruit and other crops is of much greater value economically than the honey they produce. Their importance was demonstrated when disastrous crops, particularly of fruit, occurred after 'Isle of Wight disease' practically wiped out British bees after World War I.

Bumblebees do have two advantages over honeybees. One is a longer tongue which enables them to work certain flowers, notably red clover, in which the nectar is beyond the reach of honeybees, except when unusual circumstances have raised the level above normal. When red clover was introduced into New Zealand there was no native insect able to pollinate it, so it did not produce seed until bumblebees were imported and established. The tongue of the honeybee is about 6mm long, a little more in some races, and even if the head can be partially inserted, the limit of its reach is about 7mm. However, if the tip of its tongue will just touch the top of the nectar, it is able to remove it all. Bumblebee tongues vary with the species: *Bombus terrestris* has a tongue 8.2mm long, and *B hortorum's* is 13.7mm. The bumblebee's other advantage is its ability to work in cold, wet weather which honeybees cannot endure.

In a bad spring, bumblebees could probably save the fruit crop if there were enough of them. It has sometimes been found for instance that despite bad weather at blossomtime, the outer trees of large orchards have produced a fair crop due to pollination by wild bees from surrounding rough ground. However, modern land usage tends to eliminate waste ground, heaths and hedgerows in which bumblebees nest, and large acreages of orchard will contain no habitat suitable for them. In nature there is a proper balance between flora and pollinating insects which man upsets when he clears and plants huge acreages—in some countries, square miles—with a single crop. This imbalance can be redressed by importing hive-bees at blossomtime—always providing that weather conditions allow them to work. In fact, even when wind and rain have stripped the petals from the flowers, honeybees will still pollinate fruit trees when conditions improve, so long as the essential parts of the flower remain. After pollination has taken place, the bees would of course have to be moved to an area containing different forage, or they would starve. A problem arises with red clover, which depends on bumblebees for its pollination, for if it is grown in huge blocks there will be no habitat for wild bees, and this species does not adapt readily to domestication. Even if colonies were established artificially, they would need alternative forage while building up to maximum strength ready for the clover's flowering.

Honeybees may go out expressly to collect pollen, especially when intensive brood-rearing demands it; but it is also collected, inadvertently or as a sideline, by nectar-gatherers, for the bee's hairy pelt picks it up as she enters the flower. The bee may pause on the flower to brush it off and pack it in her pollen baskets, or may do so while flying between plants. Sometimes a bee may be seen dangling from a flower by one front leg while the other five are all busily engaged in sweeping up. However, there are times when bees in flight can be seen deliberately ridding themselves of pollen, as if aware that it is not needed, or not of the right kind. For some reason, bees seem to dislike the pollen of lime blossoms and seldom take it back to the hive if it is picked up while collecting nectar.

Nectar is produced by groups of special secretory cells called nectaries, usually situated in the base of the flower, but sometimes elsewhere; they are found in the sepals of lime flowers, for instance. There are also extra-floral nectaries on the young leaves of some plants—the common laurel is

35 Stamens of the apple blossom form a tube which protects the nectar

one—where bees, wasps and flies may sometimes be seen feeding, though from the plant's point of view these would not seem to serve any useful purpose. Nectaries are often protected by stiff hairs or some other device, or by being deep down, so that short-tongued insects like flies, which are not such good pollinators, have difficulty in reaching them. Others, such as those of ivy and the umbelliferous plants like cow-parsley and fennel, are completely exposed so that many kinds of insect may be seen feeding on the inflorescence. Some flowers are hooded or pendant, which protects the nectar from dilution by rain. Many flowers have nectaries of a different colour to attract the bee's attention: other nectar-guides take the form of streaks, spots or veining which contrasts with the ground colour of the petals.

Tests have shown that bees do not see colours in the same way as human beings do. They are colour-blind to red, which probably appears to them as some shade of blue, and they do not distinguish between yellow, orange and grass-green, nor between blue and various shades of purple. Their eyes are, however, sensitive to ultra-violet, which human eyes cannot see. Two white objects, one of which reflects ultra-violet while the other does not, will appear as different colours to the bee: the white which does not reflect ultra-violet is seen as blue-green. Therefore there are four colours which bees can clearly distinguish; yellow to green; blue-green; blue to purple; and ultra-violet.

Borage, a good bee-plant

The majority of bee plants fall within the blue to purple

36 The dandelion flower is a valuable source of pollen

range (consider thyme, borage, lavender, sage, *Phacelia*) or are yellow (dandelions, rape, mustard). The nectar-guides on petals are in contrasting colours which bees can distinguish, very often yellow or white on blue, but sometimes they are darker stripes. Examples are the yellow on blue of forget-me-not, the purple streaks on yellow pansy petals, and the stripes on germander speedwell. Some flowers have areas with ultra-violet reflection and others without, which form patterns visible to the bee though not to human eyes: these also act as nectar-guides.

The amount of nectar secreted by a flower varies with soil, rainfall, temperature, age of plant and time of day.

37 Bee on white clover; the bristles which form the pollen-basket are clearly shown

Buckwheat, which is grown in huge quantities in parts of Russia and America and yields a dark, rather strongly flavoured honey, is rich in nectar in the morning but later stops producing it and is ignored by bees in the afternoon. Bees prefer flowers in direct sunlight to similar ones in the shade, which is no doubt due to a difference in nectar levels or concentration. Others are only worked in years when conditions ensure abundant secretion. The false acacia or locust (*Robinia pseudoacacia*) is a rich and reliable honey tree in warmer climates but in Britain is only visited by bees in an exceptionally hot summer. This may also explain why hawthorn is well worked some years and ignored in others, though it might also be the result of its flowering coinciding—or not—with that of some preferred forage plant. Some flowers, such as white clover, are said to produce nectar best when hot days follow cool nights: others—lime is one—prefer moist warm days and nights. Bright dry weather is not ideal for honey-producing as a lack of rainfall definitely inhibits nectar-secretion; and shallow-rooted plants like white clover suffer badly in a drought. Bees are clever about discovering which flowers are at their peak of production and are quite capable of working one plant early in the morning, switching to another during the hours around midday, and reverting to the first as the afternoon cools. Nectar may be watery early in the day, but evaporation (especially from an open type of flower) may increase its concentration later and make it more acceptable.

Whatever the flower, the anthers bearing the pollen are situated so that the visiting bee will be sure to brush against them. Pollen will adhere to her fluffy coat and be carried to the next flower on her itinerary. Some of it may be deposited on the stigmas of the same flower as she passes, or some of the pollen brought from a previous flower may be left on them. Some plants, such as poppy and wood anemone, are what is called self-incompatible, and cannot be fertilised by their own pollen. In others, their own pollen grains send tubes down through the style at a much slower rate than pollen from another flower, so self-fertilisation is only a last resort. In fact, many plants have arrangements to prevent self-fertilisation. Cucumbers have separate male and female flowers on the same plant, one kind containing stamens and the other a style and stigmas. Most hollies have male and female flowers on separate trees so two are necessary to produce a crop of berries: the bees obtain nectar from female trees, nectar and pollen from male ones. Buckwheat is dimorphic, that is, there are two kinds of flowers borne on different plants but both possessing male and female organs. One sort has long stamens and short styles; the others have long styles and short stamens. As pollen is picked up from the different flowers on different parts of the bees' bodies, so it is transferred to the opposite parts of flowers with the reverse arrangement. Both kinds of flower produce seed; and seed from either kind will produce plants of both forms. Other plants have parts which mature at different times. The young flower of the rosebay willowherb has a style which is tucked out of the way while the anthers are producing the pollen which will be carried to other flowers by foraging bees. Later the anthers shrivel and the style stands up with its stigmas receptive to any pollen which may be brought by a bee coming from a younger flower. In a blackberry flower, the outermost anthers produce pollen first, then the next ring, but if the stigmas have still not received insect-borne pollen from

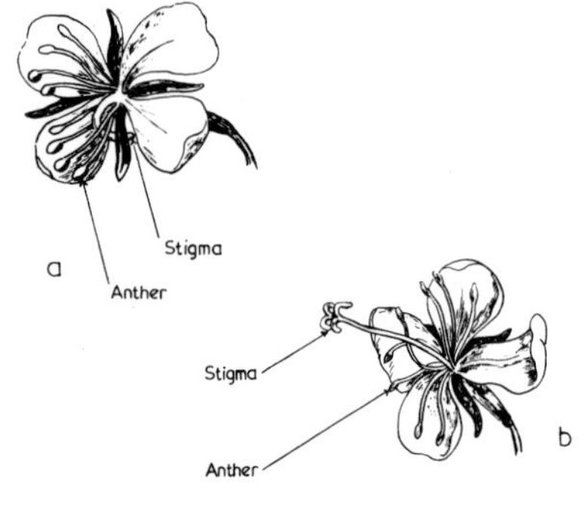

Pollination of rosebay willowherb: (a) young flower, (b) older flower

38 Worker collecting nectar from disc florets of Michaelmas daisy

another flower, the last central anthers to ripen will pollinate them.

However, the best laid plans may go awry, and bees do find ways of bypassing the ingenious pollinating mechanisms which they would encounter if they worked the flowers as nature intended. Bumblebees have stronger mandibles than honeybees and often bite holes through the base of a flower so that nectar can be removed without touching the reproductive parts at all. Honeybees are quick to take advantage of the holes provided. Runner bean flowers, comfrey, *Wiegela,* fuchsias and even daffodils are treated this way; and so too are the blooms of the antirrhinum or snapdragon, which have mouths so tightly closed that only the burly bumblebees can thrust them open, and they often prefer to take a short cut. In flowers of the *Brassica* or cabbage family there is enough space between the sepals and bases of the petals for honeybees, approaching from behind, to insert their tongues and extract the nectar while avoiding their duty. I have watched them do this to bluebells too.

It is almost impossible to carry out this sort of evasion when the parts of the flower are fused into a tube and a number of tubes are packed tightly together to form an inflorescence as in clover or teasel, or the daisylike heads of the *Compositae* family. Each floret of these heads is a complete flower. In white clover, the outer florets mature first and, once pollinated, turn brown and drop down while the bees work the next ring. There are two kinds of floret in a Michaelmas daisy, the disc florets which are packed together at the centre and the surrounding ray florets which carry the conspicuous petals. Some of the composites, such as thistles, have heads consisting entirely of tubular florets; in others, dandelion for instance, all florets are ligulate, ie have a ray. Quite often the conspicuous outer florets of a composite flower are sterile, and double forms of these flowers are invariably sterile. As the reproductive parts have been modified to form the additional petals, bees do not work double flowers, and the plants must be propagated vegetatively. Ornamental double-flowered cherries and crab-apples do not produce fruit. While bees work the older simpler forms of marigold, hollyhock, Michaelmas daisy and other garden flowers, the more spectacular multi-petalled forms developed by nurserymen are of no use to them.

The amount of pollen one colony of honeybees needs in one year has been estimated as between 50lb and 75lb, every grain of which must be carried into the hive by the workers. It exists in huge quantities, for instance a single apple blossom may produce 70,000 pollen grains and a bee may carry 50,000 or more on its body at a time. Only ten grains are needed to fertilise another apple blossom. With cherries and plums, just one viable grain is necessary to produce a fruit. There is a tremendous difference in the size of individual pollen grains, ranging from 140 microns for hollyhock down to 6 microns for forget-me-not. (A micron is $\frac{1}{1000}$ mm or $\frac{1}{25,400}$ in.) Forget-me-not pollen is often found in honey, although nectar from the flower may be a very small proportion of the whole, simply because the tiny grains are so easily swallowed by the bee when collecting nectar; and also because they evade most filters. The shapes of the various types of pollen grain are extraordinary and very diverse when examined under a microscope. Some flowers are very lavish providers of pollen—hollyhock, for instance, or dandelion. Several bees may be found in one of these blooms and all will emerge smothered with pollen so that their baskets are very rapidly filled.

To assist adhesion, pollen is moistened with nectar from the bee's honey-stomach or from the plant as it is being packed, and this alters the colour. The honeybee's load may not appear to match the actual pollen visible in the flower: it is generally darker. Observant beekeepers learn to recognise the source of pollen going into the hive, and this can be informative. It is disconcerting to find loads of bright-orange dandelion pollen being carried into hives when the bees are supposed to be pollinating apple-trees which have yellowish-green pollen. Most areas have a regular sequence of pollen plants. In Britain, it might begin with hazel, willow and blackthorn; continue with arabis and aubrietia, wild and cultivated fruits, sycamore and horse chestnut; go on to the summer harvest of clover and poppies; and wind up with an autumnal crop from dahlias, Michaelmas daisies, ling and ivy. Much pollen looks grey-green or yellow in the bees' pollen baskets but some of the colours are startling, for instance vivid orange from yellow crocus, dandelions, broom and dahlias; deep brick red from purple dead-nettle and pink-flowered horse chestnut; rich turquoise from rosebay willowherb; deep blue from *Phacelia tenacetifolia* and *Scilla siberica;* and black from field poppies. Some flowers produce no nectar but their abundant pollen attracts the bees, ensuring that fertilisation takes place: among these are the various kinds of poppy, wild and cultivated; St John's wort; mullein; and dog-rose.

Equally important to the bee is the nectar she obtains from

39 Up to the wings in pollen!

the flowers. We tend to regard as good honey-plants (in fact they are *nectar*-plants) those which man grows in large quantities because he has a use for their fruits or seeds, irrespective of any honey which may be produced from them. Fruit trees, including citrus and soft fruits; the *Leguminosae* such as field beans, sainfoin and the various types of clover; buckwheat; sunflowers, which are grown on a large scale in Russia for oil; mustard, oil-seed rape and other *Cruciferae*, all come into this category. However, a weed like dandelion

40 Honeybee with large load of wallflower pollen

would be equally valuable as a honey-producer if it were available on the same scale. Where dandelions are abundant in orchards this has sometimes been found detrimental to pollination of the fruit trees because the bees preferred the weeds. A sycamore tree in flower will roar with bees, giving the impression that a swarm is in the air, and large amounts of honey may derive from this source, but sycamores are not commercially valuable, and the millions of swift-growing seedlings distributed over the neighbourhood as a result of the

41 Thyme is a good source of both nectar and pollen

bees' activity are never appreciated. Charlock is as good a honey plant as mustard but farmers regard it as a troublesome weed and strive to eliminate it from cornfields and fallow land. Mustard intended as green manure or fodder for sheep is not in any case allowed to flower, as it does when grown for seed to make our indispensable condiment. Gardener and farmer alike wage war on bindweed, but the smaller pink-flowered *Convolvulus arvensis* is well worked by honeybees and might yield a good crop of honey if it were anywhere allowed to grow freely. A related species, the campanilla (*Ipomoea triloba*) is a notable source of honey in Cuba. The wild figwort (*Scrophularia nodosa*) is a marvellous nectar-producing plant though not otherwise useful to man, nor sufficiently ornamental for gardens; a related figwort has been grown as artificial bee-forage in the USA under the name of Simpson's honey-plant.

With costs of cultivation ever rising, and land usage becoming more specialised and intensive, there is no likelihood of crops being grown solely for bees; but in places unsuitable for ordinary agriculture, certain plants might flourish unchecked and provide valuable forage. This happens naturally: the heather which covers the moorlands of England and Scotland, and the thyme which grows freely on hillsides around the Mediterranean are examples. On a smaller scale, the rosebay willowherb which appeared magically on bombed sites in London during World War II supplied urban beekeepers with many pots of first-class honey. Waste ground along railway embankments, worked-out quarries and such places could be made equally productive, on the lines advocated by John Hill, a Covent Garden apothecary, in the eighteenth century. Thyme and origanum (marjoram), he pointed out, will grow on any dry waste ground; and lavender, which really needs cultivation, could be marketed for profit. Hyssop would also do well on dry, sunny slopes; and there are other undemanding bee-plants, such as the handsome Viper's bugloss *(Echium vulgare)* which is found on chalky hillsides, the white and yellow sweet clovers or melilots, and the purple mallow which belongs to the hollyhock family. Thrift would spread on salt marshes near the sea; and damp, waste ground near rivers would bear useful crops of purple loosestrife, various mints including water mint and pennyroyal, marsh marigolds and teasels. One variety of teasel *(Dipsacus fullonum)* has been grown as a field crop because the spiny heads were used—and still may be, for all I know—to raise the nap on cloth. The wild teasel *(D sylvestris)* is similar but has straight spines instead of hooked ones. As the small tubular flowers open in succession from the outside, they provide nectar for several weeks.

It is seldom realised what a vital role wild plants, and the flowers in small gardens, play in the economy of the hive. There are periods of especially plentiful forage which are known as 'honey-flows', such as occur when the cherry or apple orchards, or the rape fields come into bloom. The colonies of bees are built up to full strength to deal with this explosion of blossom, but it is of fairly short duration; and unless the bees are to eat the honey they have just stored, and then die of starvation, some other crop must be available. Should sycamores abound, another honey-flow may follow the fruit blossom, otherwise the bees may have a thin time until, for instance, white clover comes into flower. The colony, at its peak of brood production, must exploit every minor source of nectar and pollen to avoid depleting its stores to a dangerous level. This means the worker bees must give assiduous attention to the holly, dog-roses and white bryony in the hedges; the dandelions, dead-nettles and charlock by the roadside; the raspberries and gone-to-flower broccoli in gardens and allotments. Times of dearth occur at different periods in different regions, according to the crops normally grown there, but the 'June gap' is notorious to many British beekeepers. With the high standard of weed control on agricultural land, the removal of hedges and the widespread policy of spraying herbicides on roadside verges, many useful minor bee plants may be eliminated. The clearing of an area of waste land hitherto covered with gorse or brambles may leave an awkward gap in the district's chain of nectar sources. At the same time, a garden may be bursting with showy plants in full bloom, but if they are multi-petalled roses and spectacular double-flowered hybrids, there may be nothing at all to fill a bee's honey-stomach.

It is to supply such seasonal deficiency, rather than in the expectation of increasing their honey harvest by any noticeable amount, that beekeepers plant bee trees and shrubs around their apiaries. Early in the year, when the bees dare not fly far from home because the weather is unreliable, a hedge of hazels or a couple of willow trees near the hives will ensure pollen at a time when it is essential. A row of lavender bushes and a well-grown specimen of the orange-flowered *Buddleia globosa* may tide hungry bees over the June gap, while a bed of Michaelmas daisies and single dahlias—or even a mass of flowering ivy on an old shed—may make a real difference to the stores with which the bees settle down for the winter. Many beekeepers—and others who like to see and hear bees at work in their gardens—grow some of the flowers which always attract them. In addition to those already mentioned, the yellow and white Custard and Cream *(Limnanthes douglasii)*, the waxy-leafed ice plant *(Sedum spectabile)*, mauve-flowered *Phacelia tenacetifolia*, mignonette, and the garden balsam *(Impatiens royaleii)* are outstanding. Shrubs

which are favoured include purple-flowered *Hebe* (shrubby veronica), the mothball bush *(Symphoricarpus albus)* with tiny flowers and white waxen berries, and various cotoneasters, notably the 'fishbone' one called *C horizontalis.* Even in the kitchen garden, bees sometimes find a bonus in the shape of

42 (left) A popular bee-plant in gardens, *Cotoneaster horizontalis*

Custard and Cream, *Limnanthes douglasii*

gone-to-flower vegetables, for turnips, carrots, leeks and onions, as well as all the cabbage family, will attract them—as happens when these same plants are grown on a large scale expressly for seed production.

6 Harvest of Honey

Honey is a unique food, which can only be produced by bees. There have been attempts to make it artificially, and something which bears a superficial resemblance to it in colour, texture and calorie value can be manufactured from sugar-cane. But it is not honey. It lacks the flavour, the aroma, and all the health-giving and extraordinary properties of honey. Only bees possess the secret of transforming nectar from a diversity of flowers into the amazing product which has sustained and delighted mankind from the dawn of history.

Nectar, as collected from the nectaries of flowers, is a watery sugar solution containing small quantities of mineral salts, mainly phosphates; gums; aromatic oils; and other minor constituents such as fat and albumen. The sugar present is mainly sucrose, the same as is found in sugar-cane and sugar-beet. The bee inserts her proboscis, or tongue, into the flower and extracts the nectar which is stored in her honey-stomach for transport to the hive. The honey-stomach of the bee is situated in front of her own personal stomach, and she is able to regurgitate the nectar taken into it on arrival home. A valve prevents the return of nectar which has passed into the next stomach to nourish her. During her homeward flight, and afterwards when the thickening nectar is being manipulated by the household bees, an enzyme called invertase is added to it, so that the sucrose is 'inverted', ie converted into dextrose (glucose or grape sugar) and levulose (fructose or fruit sugar). Heat assists the process. While these changes are taking place, excess water is being evaporated from the semi-processed honey in the warm well ventilated atmosphere of the hive. Nectar from different plants contains water in different proportions: nectar from pear blossoms is very dilute, while in hot dry weather the nectar in the flowers of the lime tree is so concentrated that actual crystals may be visible. An average nectar contains about 60 per cent water, but weather conditions affect the concentration, and there are often variations according to the time of day.

When the correct density has been achieved, the bees cover the honey with wax caps. Ripe honey will keep for a very

43 One of the last skep-makers. 'Granny' Cooper, aged 92 (in 1961), New Forest, Hampshire *Robin Fletcher*

long time, but honey taken from the hive before the bees have capped it is liable to ferment, due to the yeasts naturally present in it. Extracted honey which is not kept in an airtight container will also ferment because it is hygroscopic, ie it attracts water to itself out of the atmosphere, and once sufficiently dilute, the yeasts develop. Good honey contains about 87 per cent of sugars and minor constituents and 13 per cent of water, but there can be up to 19 per cent of water before there is much risk of fermentation. The sugars, dextrose and levulose, occur in almost equal proportions, usually slightly more of the latter, plus about 1 per cent sucrose and small amounts of some more complex sugars. Pollen grains are always present, however thoroughly the honey is filtered; the richer the pollen content, the better the flavour is, and the more protein, amino-acids and vitamins the honey will contain. The food value averages 1,380 calories to the pound.

Honey may be of any colour between palest gold and very dark brown. The flowers from which the nectar came determine the colour, as well as the flavour and fragrance. Honey from rosebay willowherb is very pale, and white clover yields a famous light honey, while darker honeys come mainly from tree blossoms and buckwheat. In a small country like Britain, where huge acreages of single crops are not grown, honey tends to be multifloral, that is, a mixture of nectars from many sources blended by the bees. Although it often contains a small proportion of gorse and other nectars, heather honey is an almost unifloral product, for beekeepers take their hives to the moorlands of Scotland, the West Country and Yorkshire so that the bees can gather a late crop from the ling (*Calluna vulgaris*). This wild plant covers large areas of western Europe but does not occur in Australia, America or Asia. Ling honey is dark and strongly flavoured, with a relatively high protein content, and is unusual in being thixotropic, that is, of a jelly-like consistency. This means it cannot be extracted in the normal way, so it is either sold in the comb, or forced out with a heather honey press which of course destroys the comb. This, and the limited quantity harvested each year, tends to make it expensive. There is also a device whereby a number of pins enter the cells and liquefy the contents long enough for them to be spun out in a centrifugal extractor. Like thixotropic paint, heather honey can be made fluid by stirring but will soon set to a jelly again. It is unmistakeable in the jar as it is always full of silvery, trapped air-bubbles. Another kind of heather honey, that from bell heather (*Erica cinerea*) is not thixotropic and is extracted in the normal way. It also is a distinctive and highly regarded product.

In countries where single crops are grown on a huge scale, unifloral honeys are common, for instance, orange blossom honey from Spain and Israel, eucalyptus from Australia, buckwheat from America. The most widely distributed unifloral honey is white clover because this plant occurs in huge quantities in the spacious pasturelands of the USA, Australia, New Zealand and Argentina. Aromatic honeys derived from sage, lavender, marjoram and other herbs are produced in countries around the Mediterranean. The rosemary honey from Narbonne in France has been held in esteem for centuries. The two honeys most desired by connoisseurs of the ancient world came from Hymettus, the mountain east of Athens, and from Mount Hybla in Sicily; both were derived from wild thyme. They are still in demand today. The dark 'pine honey' which is exported from Germany—and commands a high price—is in fact honeydew, a sweet sticky liquid exuded by the aphids which infest the great conifer forests, feeding on the sap of the young shoots. This is collected and processed by the bees in the same way as honey. Conifers do not bear nectar-producing flowers, nor—as some people would like to believe—extra-floral nectaries.

44 Honeycomb sections—natural honey at its best

Honeydew is also found on sycamores, limes and other trees, and a good deal of it probably goes into hives everywhere in most years. It is as valuable nutritionally as honey, and perfectly wholesome. As anyone who has a lime or sycamore in the garden will know, the sticky honeydew generally becomes covered with a black unsightly mould, like soot, as summer advances, and this, when taken into the hive, does spoil the colour of the honey. I have tasted pure honeydew 'honeys' which looked like axle-grease but were palatable, and in some cases excellent.

Pure honey, if it is not heat-treated, will in due course granulate or crystallise. How soon, and with what texture, depends on the source of the predominant nectar and the temperature at which it is stored. The dextrose forms crystals first, followed later by the levulose, so that honeys with a high proportion of the latter sometimes remain liquid with clusters of crystals suspended in them. This looks odd and deters people from buying it, though labels explaining granulation are occasionally put on the jars. Honey derived from dandelions, oil-seed rape, or any of the *Brassica* family (including run-to-flower cabbage and broccoli) may crystallise in two to three weeks, often while still in the comb. Others, such as that from *Robinia,* may remain liquid for years. Frosting is another thing which worries customers, though it should not. This silvery patterning against the glass of the jar occurs in high-quality honey with a predominant dextrose content. It is in fact an uneven crystal formation caused by air leaving the liquid matrix of the granulated honey in cold weather, and is a guarantee that the honey has not been heat-treated, and was probably bottled straight from the extractor.

Honey in the comb is as nature and the bees meant it to be, full of flavour and aroma. Anything done to such a delicately balanced product is likely to alter it, and in most cases degrade

45 Romano-British honey jar, AD150-200 *IBRA Collection*

its quality. The small beekeeper who is careful about cleanliness, and who bottles his honey soon after extraction, with the minimum of straining, is likely to produce the least spoiled honey. Large-scale commercial producers store honey in bulk, and foreign honeys are imported in large containers; both must then be reliquefied for bottling. Whether it is heated at a low temperature for a long time or at a high one for a shorter period, enzymes and vitamins are destroyed, and the aroma—and often the flavour—suffers. Honey heated to 60°C (140°F) will not granulate again unless it is seeded with a small quantity of crystallised honey, when it will set with a similar crystal. This treatment is therefore given to honey which crystallises with a naturally coarse grain (as *Brassica* honeys do) to give it the more acceptable fine texture.

The pollen grains present in all honey give it a cloudy appearance. To produce a clear, even brilliant, liquid honey, it is filtered in various ways. At one end of the scale, a mesh which screens out particles of wax and the coarser pollen grains scarcely affects the honey; but the use of fine filters, and in particular a filter press, destroys much of its value in the pursuit of 'eye-appeal'. Honeypacking is a commercial business, and the product cannot be handled as the hobby beekeeper handles his, but over-filtering, and over-heating to ensure that the honey flows easily and rapidly through tanks, filters and pipelines, cannot but affect those properties for which many people buy honey, and which they have a right to expect in it. Unfortunately, appearance is the basis on which most purchasers decide between one honey and another. It is remarkable that the many centenarians whose longevity appears to be due to eating honey—and in some places there are villages famous for the extraordinary age and vigour of their 'senior citizens'—always belong to people who practise a rather primitive type of beekeeping. Their honey is an unrefined product, just as it comes from the bees, and would probably not be regarded as saleable among the more 'developed' nations. It would be interesting to know if there are any centenarians who attribute their age to shop-bought honey.

What are these properties which have so long been acclaimed as miraculous? Such a fog of folklore and superstition envelops the subject that it is difficult to discover the truth. In the past magic and medicine were almost synonymous, which further complicates the issue, because curative claims for honey were attributed to supernatural forces—the bee being itself a sacred and marvellous creature—as much as to any virtue intrinsic in the substance. Ill-health was generally regarded as due to bad luck rather than to any defect or malfunctioning of the body, so a spell or charm would be an appropriate treatment. A knife-wound could be seen and, therefore, treated with healing salves; a headache or stomach ache—or for that matter epilepsy or TB—had no visible cause and could be ascribed to witchcraft. This was the attitude for hundreds of years during which physician and magician meant almost the same thing. Honey given as part of religious and superstitious ritual probably effected cures for which the mumbo-jumbo received the credit.

At the same time, purely medical claims were made for honey by scholars of ancient times. Some were very odd. What can one make of the statement in *De Mirabilibus Auscultationibus,* written in the third century BC, that honey from box plants would drive sane people mad but cure epilepsy? It is debatable which would be the worse handicap. It was also thought that honey from the common *Rhododendron ponticum* or *Azalea pontica* was poisonous. Xenophon (born 430 BC) wrote that a Greek army of 10,000 soldiers retreating from Persia ate honeycombs from hives in a village near Trapezus, in an area where these shrubs abound, and were afflicted with such vomiting and purging that none could stand. None died, however, and recovery was complete in three days. Pots containing this Pontic honey, according to Strabo some 400 years later, were actually put in the road for Pompey's cohorts to find, so that the troopers could be easily killed while helpless from its effects. Not surprisingly, in view of the unfortunate characteristics of their local product, the inhabitants were required to pay their tax in wax instead of honey. This laxative effect seems to be fact, since even in England people claim to have suffered after eating honey derived from the common rhododendron, which grows abundantly in some woodlands. However, the flowers are worked mainly by bumblebees. In 1790, people in Philadelphia, USA, were supposed to have died after eating honey from *Kalmia latifolia,* a shrub related to rhododendron. It is unlikely that it grows anywhere in sufficient density to make it a hazard. No other honey is considered harmful, though nectar from ragwort and privet imparts an unpleasant flavour. The nectar from some species of lime, notably *Tilia petiolaris,* the weeping silver-leafed lime, has a stupefying effect on bees, which may fall to the ground in thousands. If not overtaken by cold or rain-showers, nor attacked by predatory birds and insects, the majority of these bees would recover, but they are very vulnerable while helpless. Bumblebees, perhaps because of the larger loads they take on, are particularly badly affected. Lime honey has, of course, no narcotic effect on human beings.

The most stupendous claim made for honey was that it conferred immortality and was the food of the gods. Nectar and ambrosia are generally deemed to mean liquid and solid

46 Bees have inspired attractive stamp designs all over the world
IBRA Collection

honey, or possibly mead (a far older drink than wine) and thick honey. Early authorities believed that though bees collected it from flowers and leaves, honey actually dropped from the air as dew. Pliny could not decide whether it was the sweat of the sky, the saliva of the stars, or a juice formed from the air as it cleared itself. Whatever it was, it had an ethereal origin, and would naturally be possessed of miraculous virtues.

On one level it conferred protection against evil, so it was used as a sacrificial or propitiatory offering to the gods, and also to consecrate ground on which temples were to be erected. It was poured over the foundations of Babylon when it was rebuilt in 682 BC by Esarhaddon. It was effective in exorcising evil demons of all sorts, and was given to new-born babies to strengthen their precarious hold on life. Even in the early Christian church, people partook of honey and milk just after they had been baptised and 'reborn in Christ'; the custom died out in the seventh century. An old Friesian law decreed that a father had a right to kill or expose an unwanted child, but not if the babe had already been given milk and honey; after that its life was sacred. Honey had a part in all the solemn occasions of life, at birth as we have seen, but also in marriage ceremonies and the rituals of death. In places as far apart as Finland and Greece, and also in Turkey, Croatia and Albania, a bride would smear the doorposts of her new home to keep away strife and give protection to the family; and in all the ancient Mediterranean civilisations, honey was one of the three (with oil, and wine or milk) traditional libations for the dead. Honey was a 'life-substance'; the food of immortality. The Hebrew's Promised Land was one flowing with milk and honey (the only two natural products intended solely as food). Rivers of honey ran through Paradise and the Isles of the Blest; and Valhalla, if not flowing with honey, was almost awash with the honey-drink, mead.

This was the mystical and religious aspect of honey, but its virtue as an antidote to evil had a more mundane application at the medical level. The Talmud said it was a remedy for gout and heart-trouble and would heal the wounds of men and beasts. It had a great reputation in Ancient Greece as a cure for eye ailments. The first-century physician, Dioscorides, recommended honey for several complaints in his book *De Materia Medica,* which was still read and studied all over Europe in the seventeenth century. The Greek philosophers Pythagoras and Democritus both lived to a great age, and both gave honey the credit. A ninety-eight year old beekeeper of my acquaintance, still incredibly fit and active, gives as his recipe 'honey and hard work'.

The teaching of the ancients was accepted as irrefutable throughout Europe in the Middle Ages, but when the new era began with Charles Butler, the medical reputation of honey was undimmed. In fact there were few ailments it would not cure, if the claims made for it could be believed. In *The Feminine Monarchie,* Butler said that honey improved appetite, was laxative and diuretic, 'nourisheth very much', prolonged life, preserved natural heat, 'breedeth good bloud', dealt effectively with ulcers, bites from serpents or mad dogs, poisoning from mushrooms and poppies, lung ailments, falling sickness, 'griefes of the jawes' and other misfortunes. Butler also said it kept the bodies of the dead from corrupting, which must have come from ancient lore as an English country clergyman surely never tested it. The honey-loving Democritus was buried in it, and so was Alexander the Great (on his own instructions), and it was a common practice in Assyria and Sumeria. The body of Agesipolis, King of Sparta, was brought back from Asia packed in honey. A gruesome tale is told of some tomb robbers searching graves near the Pyramids who came across a huge sealed jar of honey which they tasted and found to be in good condition; further down they discovered the well-preserved body of a child. It seems that at one time in Burma, funeral preparations took so long that bodies were stored in honey in the meantime; it was thriftily scraped off before the actual ceremony took place and sold in the market. A macabre thought, but honey was regarded as incorruptible.

Its virtue of resisting putrefaction made honey a valuable salve for septic wounds and old sores. Butler also explained that a 'quintessence of honey' obtained by distilling would dissolve gold and precious stones and make them drinkable, as well as reviving anyone who was dying. Could he have believed this? Such fantastic claims must have thrown doubt on the more reasonable ones.

In the eighteenth century, a small book called *The Virtues of Honey* by the Covent Garden apothecary, John Hill, was published. It is very practical, giving precise instructions for making up the remedies, and advising which honey is best for each. Hill gives honey credit for curing asthma, the gravel, coughs, hoarseness, a 'tough morning phlegm' and consumption, and adds that regular use of it would prevent many disorders. He mentions that much honey was being imported at the time, and that this would be unnecessary if people bothered to keep bees and planted crops for them. He speaks of honey in the shops being adulterated with flour and other ingredients, and therefore useless as medicine.

Belief in the curative properties of honey has persisted in all parts of the world to the present day. My mother knew how to make a syrup of honey and hyssop which had long been used in her family for coughs and bronchial troubles, and our old nanny regarded a daily spoonful of honey as an excellent restorative for convalescents and 'lackadaisical' children. Many people who do not like honey will keep a pot handy in the winter 'in case of colds'. The mixture of one teaspoonful of glycerine, the juice of half a lemon, warm water and honey is made in many households to ease sore throats, despite the various kinds of linctus on sale in any chemist's.

What is the truth about the value of honey, shorn of superstition and myth? It is a very quick-acting source of energy, as dextrose is absorbed directly into the bloodstream without further digestion, and levulose after only slight modification. Athletes find it particularly helpful as the charges of energy are thus released in sequence. The Egyptians, who made it part of the official rations issued to the king's messengers, knew exactly what they were doing. This release of energy probably gave honey its otherwise undeserved reputation as an aphrodisiac. The word honeymoon does not seem to have derived from this idea, however, but from the month of mead-drinking with which Viking nuptials were celebrated. As a quick 'pick-me-up', honey is useful for diabetics and in cases of fatigue or debility. This probably explains why Butler thought it promoted appetite—if one were too tired to eat, or had been long fasting, a spoonful of honey would provide the energy to tackle something more solid. Being predigested, it is helpful in digestive upsets, and even when the stomach is ulcerated, for honey's antibiotic property is now established.

Since it inhibits the growth of bacteria, and is therefore a natural disinfectant, honey's use for dressing wounds and septic ulcers is endorsed by science. (Possibly this also explains why honey from corpses could apparently be eaten without ill-effects.) The antibiotic function seems to be contained in a substance called 'inhibine' which some researchers think may be hydrogen peroxide, a breakdown product of glucose with well known bactericidal properties. This constituent would not survive heat-treatment, nor probably would any other anti-bacterial substances which might be present. Hence the old cottager, treating an injury with honey from his own hive, could bring about an improvement; but the buyer of imported honey from a health food shop would not achieve the same result. The preservative quality is valuable and honey is included in some medicines just for that reason. It is also used in bread, which keeps fresh longer, and in curing hams.

The slightly laxative property makes honey an ideal corrective for children and old people. Other claims which stand up to scientific investigation are that it 'breedeth good bloud'—it very readily produces haemoglobin in blood—and that it strengthens the heart. It is in fact quite a powerful stimulant and will often help the sufferer from a heart-attack. Perhaps in those far-off days when it was given to newborn babies for religious reasons, it strengthened their hold on life in a physical way as well as a spiritual one.

Although a doubtful protection against witch-craft or serpent bites, the curative value of honey is in many respects well established. Still controversial, however, is its use as a treatment for rheumatism, in the form of cider vinegar or oxymel. (The BPC formula is: acetic acid, 150 ml; purified water, 150 ml; purified honey to 1,000 ml.—Dose: 2.5-10 ml.) Miraculous results have been claimed for it, but many people are unconvinced. Rheumatism is not one disease, however, but many. Some forms may respond to this treatment while others do not. The same thing may be said of the ancient belief that bee stings, or more accurately bee venom, will cure rheumatism. Some beekeepers are crippled with it, despite lifelong injections of bee venom; others claim a complete cure. So far as I know, no one still treats patients with an old-fashioned 'bee-vaccinator', but at this time a clinic exists in London where a Mrs Julia Owens gives bee sting treatments, both for arthritis and for blindness caused by *retinitis pigmentosa.* In the light of some extraordinary results, the curative properties of bee venom may be due for serious, unprejudiced investigation.

Though its price is comparatively high, honey is a much healthier sweetener than sugar, and can be used in confectionery and ice-cream, sweets and stewed fruit. As it contains up to 20 per cent water, the moisture in recipes must be reduced—$\frac{1}{5}$ of a cupful less water or milk for each cupful of honey used. It is usual to allow $1\frac{1}{3}$ cupfuls of honey against each cupful of sugar to obtain equivalent sweetness.

There are minor uses of honey: as an ingredient in hair tonics and hand lotions, for curing tobacco, even as anti-freeze for the radiators of cars. Traditionally, apart from being used as food and medicine, large quantitites were employed in the making of those alcoholic drinks collectively known as mead. It is certain that mead was the first fermented beverage known to mankind, far older than beer or wine. The Maya made it from the honey of the little stingless bees (*Meliponinae*), as did the Aborigines of Australia up to modern times. East African tribesmen make a fermented honey-drink now and a similar one was known in the ancient civilisation of India. Even when wine overtook them, honey drinks were still made and consumed, as much for medical reasons as for pleasure, in ancient Greece and Rome. The words used for them are confusing. *Mulsum* was a mixture of wine and honey, but not true mead because the honey itself was not fermented. It had a good reputation: for instance the centenarian, Pollio Romulus, told the Emperor Augustus that his excellent health was due to mulsum used internally and olive oil externally. Columella said mulsum was made of $\frac{4}{5}$ wine and $\frac{1}{5}$ honey.

More truly mead was the beverage known as hydromel or *aqua mulsa.* Pliny's recipe for this was to add $\frac{1}{3}$ of honey to rainwater and keep it in the sun for forty days at the rising of the Dog Star. He claimed that 'new' hydromel was good for the sick, for people liable to take cold, and for small-minded persons! It healed the temper as well as the body, aided respiration, soothed coughs, and acted as an emetic when warm. Palladius mentions exotic variations such as rhodomel, made from roses and honey; and omphacomel, made from grape juice and honey. Pliny describes oxymel which was made from honey, vinegar, rainwater and sea-salt—not quite the same as the BPC formula of the same name—a reputed cure for throat and ear troubles.

However, it is the Vikings with their foaming horns of strong drink, berserk rages and blood-lust who come most readily to mind when mead is mentioned, and we tend to forget that it was drunk all over Europe before the invention of beer. Mead was the drink of the ancient Irish gods; rivers of it flowed through the Celtic Paradise; and it was the tipple consumed in the great banqueting-hall of Tara. If it had not been known before, the Romans would certainly have introduced it to Britain; and the Saxon invaders who came later were people with a still greater addiction. The Norse sagas are saturated with mead—Thor was supposed to have drunk three tuns on one occasion—and the dead heroes who went to Valhalla could indulge to their limit. Just as honey gave magical gifts amongst Mediterranean peoples, so mead was reputed to confer the blessings of poetry and oratory, wisdom and immortality on the chillier northern races which clearly needed a drop of strong drink to loosen their silver tongues.

In the Middle Ages in Britain, wine (mostly imported) was drunk by the very small aristocracy, beer was sold in the alehouses, but mead was a drink to be made and consumed in all ordinary homes. There were as many varieties of honey drink as of home-made wine. The twelfth-century abbot, Alexander Neckham, writes of clare and piment, but these were mixtures of respectively white and red wine with honey and spices. Morat, a fermented drink made from mulberries and honey, was a sort of fruit mead. Another variation, popular in Wales, had malt and spices added and went under the name of braggot. True mead was held to contain only honey and water; the same brew flavoured with herbs and spices was metheglin—something of which Queen Elizabeth I was particularly fond. She had it made especially for her from a recipe which included sweetbriar and bay leaves, thyme, mace and cloves.

Mead continued to be drunk throughout the eighteenth and nineteenth centuries in country districts, and even today many beekeepers make it. People who have to buy honey probably regard as gross extravagance the use of three or four pounds of it to make one gallon. Some mead is made commercially and public houses sometimes stock it, rather as a gimmick. Some I sampled was horrible, but the landlord confided to me that he last sold a glass from that bottle a year and a half earlier, which may have been the reason.

That could not have happened 300 years ago, when 'the Eminently Learned Sir Kenelme Digbie' sang its praises, and devoted a hundred pages to mead recipes in his book *The Closet Opened.* Of a friend who drank nothing else Sir Kenelme wrote, 'And though He were an old man, he was of extraordinary vigor every way, and had every year a Child, had always a great appetite, and good digestion; and yet was not fat'. Less has been claimed for more pretentious drinks, and fashion might yet restore it to favour.

7 Beekeeping in Early Times

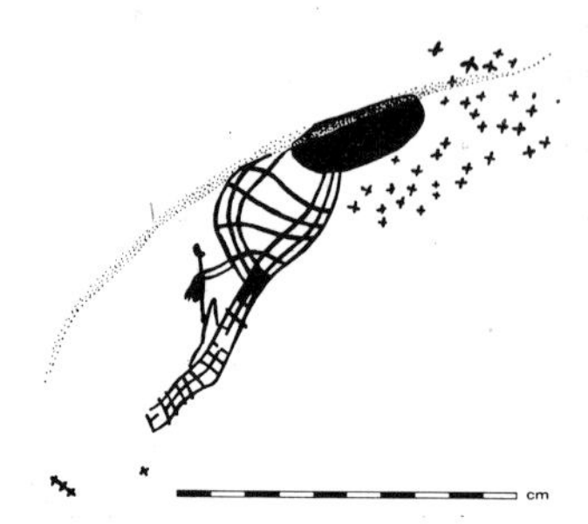

Bushman rock-painting from Eland Cave in Natal Drakensberg, showing honeygathering *Harold Pager*

Rock-painting from Matopo Hills, Rhodesia, showing bees' nest being smoked (for enlargement see p. 136) *Harold Pager*

Primitive man was a bee-hunter, not a beekeeper. He climbed trees and braved stings to get the honeycombs; and before long he discovered that smoke from a burning branch would drive off the bees and make it easier to rob the nests. He lived a wandering life, feeding off the countryside as he went, but he might remember and revisit a cave or hollow tree where he had found honey before. Bees tend to reoccupy the same sites. When a more settled way of life evolved, a tribal group would gradually build up a knowledge of many such trees in their neighbourhood. Leaving a little brood comb each time would encourage the bees to return, and the nest could be robbed at regular intervals, rather in the manner of a beekeeper visiting out-apiaries. Later on, artificial holes might be made in living trees, or hollow logs hung in the branches for swarms to occupy. This system—halfway between bee-hunting and beekeeping—was used by the *bortniks* in the great forests of Russia until well into the seventeenth century; and African tribesmen practise it still, marking the bee-trees with signs and symbols to denote ownership.

Laziness being the mother of invention, it eventually became obvious that it would be convenient to have the bees near at hand, so avoiding the dangerous journey through the forest. Different races reached this realisation at widely varying times and there are still primitive peoples today who are bee-hunters. No one knows where the very first beehive was used nor what it was like. It may be that some early agriculturalist found a fallen piece of hollow tree containing bees and dragged it nearer his home—the tools available to him would not have encouraged him to cut one down. Log-hives have persisted in Africa and Russia and in the 'bee-gums' of America until modern times. Elsewhere the idea of a home apiary may have taken root when a swarm took possession of a basket left lying near a dwelling-hut or in the forest. The hives of woven osiers daubed with clay, or dung and ashes, which were in common use over most of Europe in ancient times were similar to the baskets made impermeable with clay in which early man carried his food.

The Ancient Egyptians were capable beekeepers as long ago

as 2600 BC. Scenes carved in relief in the Temple of the Sun at Abusir show them smoking the bees and taking out the honeycombs, pouring the honey into jars and sealing them. The fuel to supply the smoke would probably be dried cow-dung, and the horizontal earthenware hives are of a type still used in some parts of the Middle East. Today, they are straight cylinders like lengths of drainpipe, but those shown in the painting of an apiary in the tomb of Pa-bu-Sa at Thebes (about 620 BC) are thicker in the middle than at the ends. One cannot help thinking that they evolved from a wine-jar or amphora. I have seen an abandoned five-gallon drum in which bees had built splendid combs; the same sort of thing may have happened thousands of years ago in Egypt and been turned to good account.

47 A log-hive from California, USA, the top of which is missing
IBRA Collection

In these long hives, the bees reared their young in the combs nearest to the entrance, using those further back for honey storage. The Egyptian beekeeper could remove the back of the hive and cut out the combs of honey, leaving the brood undisturbed; then the end would be plastered over again and the bees would replace the combs. By reversing the hives periodically, the combs could be regularly renewed. Because they had no long cold winter to survive, these bees would not store as much honey as the northern races but the season would be longer and honey could be taken more frequently. This kind of 'depriving' system, which did not involve killing the bees, was followed in all the countries around the Mediterranean where bees were regarded as semi-sacred creatures. The Egyptians also practised migratory beekeeping, moving their hives on rafts down the Nile to take advantage of the later-flowering crops nearer to the estuary.

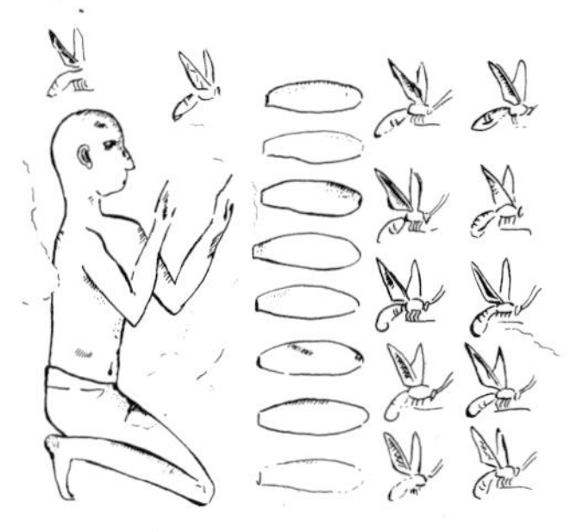

Beekeeping in Ancient Egypt (tomb of Pa-bu-Sa at Thebes, 620 BC)

In Northern Europe a different method of beekeeping developed to suit different climatic conditions. There, bees needed a large quantity of honey to keep them alive through the winter. If the beekeeper took it all they would die: even if he left them some, there was no certainty that it would suffice. It became the custom to kill most of the colonies and take all their stores, hoping that those left undisturbed would winter well and send out early swarms to replenish the stocks. The lightest and heaviest hives were sacrificed, which must have tended to perpetuate a very mediocre type of bee.

The earliest account of beekeeping still in existence occurs in the ninth book of Aristotle's *Historia Animalium* or *Natural History*. This part is now thought not to have been written by Aristotle but by a practical beekeeper living rather later, probably in the early part of the third century BC. He deals with the need to feed bees with sweet liquids if they had insufficient stores, but in his view leaving them too much honey made them lazy. He knew that worker-bees gorge when they are smoked, and die when they sting, and do

48 A Greek wattle and daub hive *IBRA Collection*

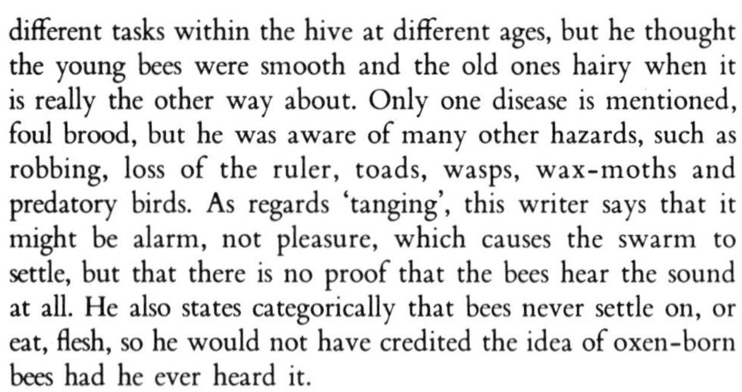

different tasks within the hive at different ages, but he thought the young bees were smooth and the old ones hairy when it is really the other way about. Only one disease is mentioned, foul brood, but he was aware of many other hazards, such as robbing, loss of the ruler, toads, wasps, wax-moths and predatory birds. As regards 'tanging', this writer says that it might be alarm, not pleasure, which causes the swarm to settle, but that there is no proof that the bees hear the sound at all. He also states categorically that bees never settle on, or eat, flesh, so he would not have credited the idea of oxen-born bees had he ever heard it.

The next books we have which deal with beekeeping were both written in Italy in the first century BC, but their authors, Varro and Vergil, approached the subject in very different ways. Vergil was born in the country near Mantua and grew up loving the simple rural life. Though he could have taken his place in the most cultured society of Rome, he preferred to live on his farm in Campania, writing poetry, keeping bees, and tending his vines and olive-trees. Seven years were devoted to the *Georgics,* a long poem on agriculture of which Book IV is concerned with bees.

Vergil begins with some sound advice on setting up an apiary. It should be out of the wind in a place where cattle and sheep will not intrude; water must be available, with sticks and stones laid in it for the bees to settle on; and a tree is desirable to provide shade and tempt swarms to cluster there. The site must be free of unpleasant smells, and echoes, and yew-trees (the honey from which he believed to be poisonous); various predatory birds and 'the spangled lizard' were to be discouraged. Flowers such as thyme, savory and violets could be planted nearby. The hives, whether of pieces of bark sewn together or woven from osiers, should have small entrances and all chinks should be stopped with clay, so that the inside was cool in summer and warm in winter. Vergil does not describe the shape of these hives, but in southern Italy they were probably recumbent cylinders, worked in the same way as the Egyptian clay ones. In the Po valley in northern Italy an upright form later became common.

He mentions the use of smoke to subdue the bees, that bees which sting lose their lives, and that they are threatened by many enemies and diseases. This leads to the recipe for restoring his stocks by use of a bullock, should the beekeeper be unfortunate enough to lose them all. On a more practical level he explains how to feed colonies with honey through hollow reeds if they become 'spiritless with hunger'; and how to check excessive swarming by removing the queen's wings (though he thought it was a king, of course). The possible advantages of clipping a queen's wings are still being debated as though the idea were new!

Sicilian or barrel hive

Vergil may have employed a slave to do the apiary work but he clearly loved and took a close interest in his bees. His contemporary, Marcus Terentius Varro, would have thought a *mellarius* indispensable, for he was concerned with the commercial possibilities of large-scale beekeeping. Varro was the greatest Roman scholar of his day, a versatile man who wrote more than seventy books on a variety of different subjects. His treatise on agriculture, *De Re Rustica,* intended to foster a love of country life and a return to the land, was written in his eightieth year. It takes the form of a discussion between several people but in its different way covers much the same ground that Vergil does. Varro is more interested in the economics of agriculture, however, for he mentions two brothers he knew in Spain who made 10,000 sesterces (about £80) a year from beekeeping on a half-acre smallholding; and a man called Seius who leased his hives for an annual rent of 5,000lb of honey. Propolis is described as an unsightly substance, but apparently the doctors in the Via Sacra were prepared to pay more for it than for wax; at this moment, propolis is again fetching high prices for medical use, after centuries of being regarded as a messy nuisance. Varro also says that the way to strengthen a weak colony is to give it a new ruler. This is 're-queening' and is correct, but the suggestion is surprising because he did not know that the ruler was female and the mother of all the bees.

Columella wrote his book on agriculture about AD 60. He was a very practical man, a retired army officer who settled on an estate outside Rome, though he was in fact a Spaniard and had lived in the country as a boy. Most of his beekeeping material was learnt from Vergil and others, but he leaves out any speculation or myths which he considers unproductive. He suggests killing one of the rulers when it is evident from its shape that a swarm contains two, and he deals with brick bee-boles (niches in walls to hold light movable hives), uniting weak colonies, and putting lamps in the apiary at night to trap wax-moths, which were a serious pest. His statement that savage bees grow tame if they are often handled is interesting. This has been taken to mean that Columella's *mellarius* was in the habit of re-queening bad-tempered colonies, since this is the only sure way of changing their temperament. However, I have noticed that bees which were vicious in an out-apiary became milder after they were moved to my garden, as though they grew accustomed to the constant presence of people: perhaps Columella's experience was similar.

This book includes an interesting description of the Roman beekeeper's tools. His smoker was an earthenware pot with a small spout at one end and a larger one—through which the beekeeper blew—at the other. It was filled with hot embers and dried cow-dung. The *culter oblonga* (oblong knife) was about 18in (46cm) in length with a wide, sharp edge on both sides and a hooked scraper for raking out dirt at the end. There was also a flat blade for cutting out honeycombs.

In the fourth century AD, Palladius wrote a book on agriculture which included some beekeeping information copied and condensed from Columella. It was arranged in a month-by-month sequence, with no space wasted on fables and theories, and was clearly intended as a text-book for the busy man. And so it proved to be. Nothing better appeared for centuries. This book was read all through the Dark Ages in Europe, and was translated into English in the fourteenth century, 1,000 years after it was written. One longs for someone in northern Europe to have written about the conditions and methods of his own time and place, but none of the people who kept bees could read and write, except the monks. The twelfth century Abbot of Cirencester, Alexander Neckham, wrote about bees in *De Naturis Rerum,* copying his information from Vergil and other sources; and an English friar called Bartholomaeus Anglicus wrote about bees in Saxony in 1250, but his authority was Aristotle. At the time, the knowledge of the ancient scholars was regarded as complete and irrefutable, so it was deemed unnecessary—if not actually blasphemous—to look further. It seems to have occurred to no one that ancient Greek and Italian methods of beekeeping had little application in northern Europe, even had monkish encyclopedias written in Latin been available to illiterate Saxon serfs.

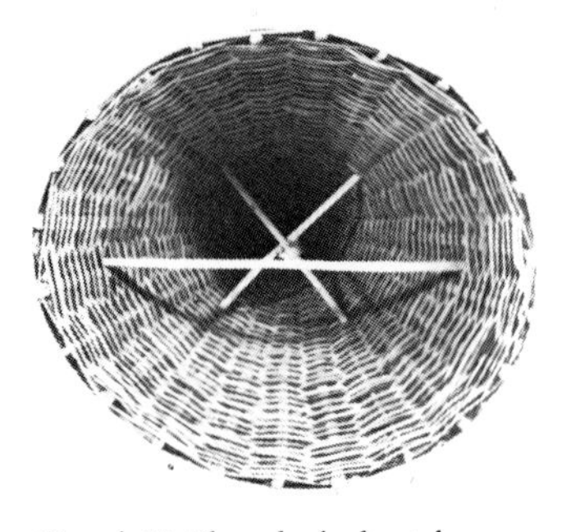

49 and 50 Clay-daubed wicker hive; (top) exterior, (bottom) interior. This came from Luxembourg but is of a type which was common in Britain for centuries *IBRA Collection*

We do know that clay-daubed upright wicker hives were used over most of Europe in pre-Christian times, and that Anglo-Saxon invaders, pushing westward through France, Holland and Scandinavia to Britain, brought with them the straw skeps which had originated in Germany. These tended to supersede hives woven of willow and hazel wands, wherever suitable straw was available. Under the feudal system the *beo-ceorl* was a lowly freeman, ranking with the swineherd. The bees in his charge did not belong to him but reverted to his lord when he died. All the forests of Britain belonged to the king, or lesser landlords, so it was poaching to take wild bees or honey from them. Most of our information about beekeeping in the Middle Ages comes from collections of old laws and records of court cases. Thus we know that King Alfred decreed that the issue of a swarm should be announced by tanging, so that it could be followed and captured; and that in 1457 a certain Richard Ruddyng of Stonehale was successfully sued by one William Mason for selling him honey contaminated with sulphur. This might indicate that bees were killed by sulphuring, but how long the practice had existed or where it originated is not known.

In Tudor times, the monasteries were dissolved. The closing of these centres of learning may have been a disaster in some ways but it had advantages. Fine houses were built because an army of skilled masons and carpenters, which had been monopolised by the religious foundations, became suddenly available. It is also possible that the release into the ordinary world of men with special training and the ability to read and write may have disseminated much closely guarded knowledge. Books too, since the invention of printing, had become slightly less rare. Several works on agriculture appeared at this time, but the sparse beekeeping pages contained nothing new. Thomas Tusser's *Five Hundred Pointes of Good Husbandrie,* written in verse about 1570, has only three references to bees. In May, the beekeeper is warned to watch out for swarms; in September, he must burn the bees and take their honey; in December, the stocks must be prepared for the winter.

Go look to thy bees, if the hive be too light,
Set water and honey, with rosemary dight;
Which set in a dish full of sticks in the hive,
From danger of famine ye save them alive.

Vergil said as much 1,600 years earlier, but at least this was in English!

In 1609 a ray of light at last penetrated the gloom in the shape of a book called *The Feminine Monarchie,* written by the Reverend Charles Butler, Vicar of Wootton St Lawrence in Hampshire. Here at last was a practical beekeeper who was also a scholar, and capable of setting down the knowledge he had gained for the benefit of others. Butler had read the ancient authorities but he did not copy what they said without question; rather he tested their statements against his experience and made his own judgements. His book is a complete guide to skep beekeeping, clearly arranged, and immensely readable.

At this time the apiary consisted of straw or wicker hives set on individual stools or ranged along a plank. The hive entrance was generally a sloping channel cut in the thick stool-top, which might be stone but was more usually wood. Sometimes a piece was cut from the rim of the skep itself, but this weakened it, and the natural compression of the straw, caused by the weight of the combs and bees, tended to constrict the entrance after a time. Sticks passed crosswise through the hive from side to side helped to support the combs which the bees fastened to them as they built downwards. In winter, extra protection was needed as a skep exposed to wet would quickly decay. Sometimes cracked earthenware pans—the sort in which cream was skimmed for

51 Woven hazel hive made in 1954 by a man who remembered them in use in Sussex *IBRA Collection*

Bee-bole containing a skep hive

A bee-house for three skeps, from the back

butter—were inverted over the top, but the traditional covering was a cone-shaped straw hackle. For this, a sheaf of straw was bound together at the top and arranged tentwise over the skep, then an iron or wooden hoop was pushed down over it to hold all secure. Field mice and spiders often took up winter quarters in them. There were also bee-boles—niches in walls—in which skeps stood all the year; and, on the East Coast particularly, the Dutch idea of a bee-house to hold several skeps was sometimes adopted, though it never became really popular.

The Reverend Butler's instructions for setting up an apiary resemble Vergil's, but he goes into great detail concerning the fences to surround it, and the necessity for keeping the grass around the hives cut short. He deals with the dimensions of hives, the height of the stools and the best way to keep apiary records. Summer and winter entrances are discussed, and instructions given for making a 'Drone-pot' with ribs at such a distance that small bees can pass in and out but drones cannot. The beekeeper is advised to leave the bees alone when they are vexed since all those which sting will die; if it is necessary to tend them, Butler says 'put a thin Cloth or Hood over your Face that you may see through, for it is the Face they most eagerly aim to strike at, and then when they strike

52 A British straw skep with split bramble binding *IBRA Collection*

the Cloth will not retain their Stings, which they can easily draw forth again.' He is always concerned for the life of what he calls 'this little painful Creature' (in the sense of painstaking, not giving pain!). Surely so humane a man must have hated killing his bees to take the honey. He gives instructions for doing it, but he also suggests driving the bees from their hive and uniting them with another stock which has plenty of stores, so that their lives can be preserved. He is clearly looking for a better way.

To take the crop, short sticks were split part-way, and pieces of paper or rag dipped in brimstone were inserted. A circular pit, about 6in (15cm) deep and the same diameter as the hive, was dug, and three or four of these sticks were planted in the bottom and lit. As the choking smoke poured up, the skep of bees was set over the pit and earth thrown up around it to seal the join. The bees tumbled dead into the pit and the skep was carried indoors.

Method of driving bees

If the bees were to be united with another stock, their hive would be turned upside down in a tub or bucket and an empty skep placed over the top. By drumming on the lower one, the bees would be persuaded to walk up into the other. For 'open' driving, the skeps were fastened together on one side with a skewer, and propped apart at the other side with the paired irons. The bees walked across at the 'hinge', and the queen could be removed en route by the sharp-eyed and nimble-fingered.

For cutting the combs from the vacated skeps, a double-ended knife (or pair of knives) was used. One end would be turned at a right-angle to the shaft and the side sharpened; this was used to cut the attachment across the top of the skep. The other was flattened and sharpened across the end so that it could be thrust up the sides of the combs to free them. These are still recognizably the tools which Columella described.

Driving irons and skep-knives could be made by any handy beekeeper or by the local blacksmith. The housewife made the linen honey-poke through which honey was strained from the mashed-up combs. An interesting version of this used today by tribesmen in Kenya is shown (Plate 53). The ingenious diagonal weave gives elasticity so that as the 'poke' is pulled lengthwise, honey is forced through the mesh.

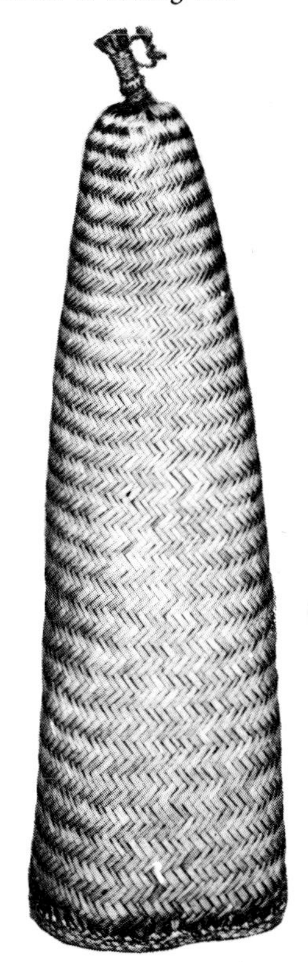

53 Honey-strainer from Kenya: the mesh expands to let honey through when the two ends are pulled apart *IBRA Collection*

The beekeeper also needed tools to make his hives. The skep-needle, to thrust through the straw when introducing the binding, was traditionally made from a goose or turkey 'drumstick'. I have seen these needles described in books and museums as bone apple-corers or marrow-scoops, which they resemble. To maintain the even thickness of the straw rope, a hollow section of a cow's horn about 2in (5cm) in diameter was used as a gauge. Sets of needle and gauge were passed

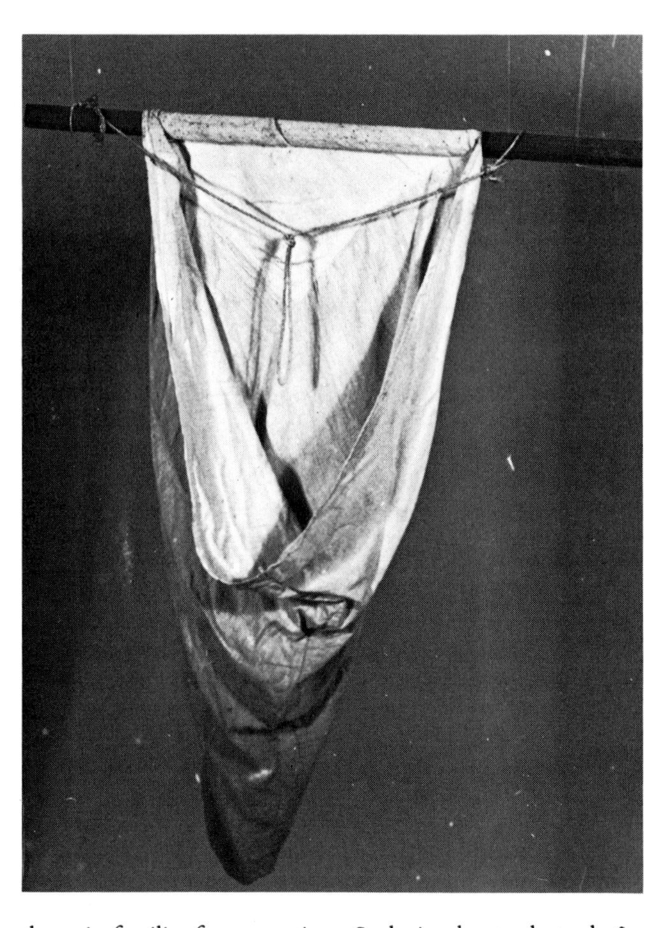

54 Linen honey-poke, used to strain honey from broken combs
IBRA Collection

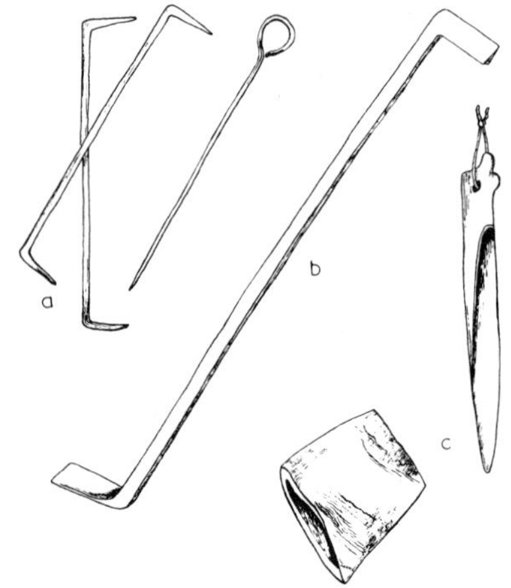

Tools for skep beekeeping:
(a) driving irons, (b) skep knife for removing combs, (c) needle and gauge for skep-making

down in families for generations. Such simple, sturdy tools are almost everlasting, and only really acquire a satisfactory smoothness after years of use. Bindings were usually peeled and split blackberry stems, and reeds and grasses as well as straw were sometimes used for skep-making. Skeps varied in shape, not only between countries (Dutch ones were always taller and narrower than the British sort) but also between regions in one country. A skep used in Norfolk or Bedfordshire was easily distinguishable from one made in Wales, or from a New Forest 'pot'. It is a pity that the differences were not recorded before the art of making them declined. A well made skep is an attractive object with its symmetrical curves (by no means easy to achieve); and though very light itself, is strong enough to support the weight of a man standing on it.

Swarming bees should be tanged, says Butler, firstly to lay public claim to them and only secondly to induce them to settle. A swarm of bees belonged to the person from whose hive it came, so long as he kept it in sight, and this still holds good. He has much sensible advice to offer about taking and hiving swarms: 'Blackberry Swarms', ie late ones, should be returned to the hive they came from, as they are unlikely to build up sufficiently to survive the winter. He knew that swarms consisted of bees of all ages, that bees worked one kind of flower at a time, that drones were male and workers female. He even knew that the ruler was a queen, and said so: this was not such an outrageous notion at a time when a

55 Memorial window to the Reverend Charles Butler in the church at Wootton St Lawrence, Hampshire

magnificent female monarch had lately sat on England's throne, but a great many people did not believe it. The great truth, that the queen was the mother of all the others, eluded him. He thought the queen bred the princesses in the queen cells, but that the workers laid the other eggs, and that they were fertilised by the drones' 'Masculine Virtue' in some mysterious way, so 'bringing forth without Coition or generating with the Male'. Everything was not known to him but Dr Butler certainly shed considerable light in dark places, and little escaped his searching enquiry. He included chapters on the products of the hive, and recipes for medicines, conserves and mead. He also had a high opinion of the bees' musical ability. In the first edition of his book he represented the hum of the swarming bees in musical notation, but later this was expanded into a four-part madrigal.

The Feminine Monarchie deservedly went into several editions, and remained in print for almost two hundred years. The Father of English beekeeping, as he is often called, was honoured on 14 November 1954, when a memorial window was unveiled in St Lawrence's Church at Wootton, where he was vicar from 1600 to 1647. A rendering of his bee-music was given by students of Worcester and Somerville Colleges, Oxford.

Before the Reverend Butler's time, beekeeping had been the province of serfs and cottagers. Some of these men may have been observant, may even have made quite revolutionary discoveries, but their ideas would have gone no further. It is difficult to appreciate how little communication there was. A man might never leave the village where he was born. He could not write, so was unable to record what he knew; and who among the small educated section of the population would have been interested in anything an illiterate peasant had to say? Whatever knowledge of his craft he acquired might be handed on in his own family; otherwise it died with him.

56 Detail of memorial window, showing inscription, and straw hives of Butler's period

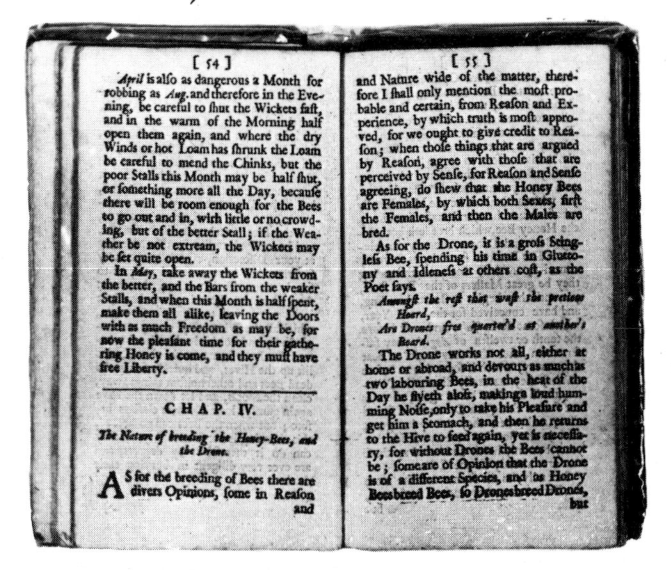

[54]

April is alſo as dangerous a Month for robbing as *Aug.* and therefore in the Evening, be careful to ſhut the Wickets faſt, and in the warm of the Morning half open them again, and where the dry Winds or hot Loam has ſhrunk the Loam be careful to mend the Chinks, but the poor Stalls this Month may be half ſhut, or ſomething more all the Day, becauſe there will be room enough for the Bees to go out and in, with little or no crowding, but of the better Stall; if the Weather be not extream, the Wickets may be ſet quite open.

In *May*, take away the Wickets from the better, and the Bars from the weaker Stalls, and when this Month is half ſpent, make them all alike, leaving the Doors with as much Freedom as may be, for now the pleaſant time for their gathering Honey is come, and they muſt have free Liberty.

CHAP. IV.

The Nature of breeding the Honey-Bees, and the Drone.

AS for the breeding of Bees there are divers Opinions, ſome in Reaſon and

[55]

and Nature wide of the matter, therefore I ſhall only mention the moſt probable and certain, from Reaſon and Experience, by which truth is moſt approved, for we ought to give credit to Reaſon; when thoſe things that are argued by Reaſon, agree with thoſe that are perceived by Senſe, for Reaſon and Senſe agreeing, do ſhew that the Honey Bees are Females, by which both Sexes, firſt the Females, and then the Males are bred.

As for the Drone, it is a groſs Stingleſs Bee, ſpending his time in Gluttony and Idleneſs at others coſt, as the Poet ſays.

Amongſt the reſt that waſt the pretious Hoard,
Are Drones free quarter'd at another's Board.

The Drone works not all, either at home or abroad, and devours as much as two labouring Bees, in the heat of the Day he flyeth aloft, making a loud humming Noiſe, only to take his Pleaſure and get him a Stomach, and then he returns to the Hive to feed again, yet is neceſſary, for without Drones the Bees cannot be; ſome are of Opinion that the Drone is of a different Species, and as Honey Bees breed Bees, ſo Drones breed Drones, but

57 Pages from a 1704 edition of *The Feminine Monarchie* by Charles Butler

The one person who could bridge the gap between the educated and the illiterate was a clergyman such as Butler, who had contact with all classes, like the monks in earlier times. A beekeeping cleric was in a good position to receive and spread knowledge, while occupying himself in a harmless and even profitable manner. Neither could the study of bees be regarded as demeaning, for had not the great minds of ancient times been fascinated by them? Moreover, their thrift, chastity, loyalty, obedience and industry supplied material for endless Sunday sermons. Throughout the seventeenth and eighteenth centuries, many books about bees were written by British clergymen, and even more in the nineteenth. It was not only in Britain that the Church and beekeeping were so involved. Clergymen such as the Polish Dr Dzierzon, the French Abbé Collin, and America's Reverend Langstroth were responsible for discoveries which revolutionised the craft.

Apart from the great names, there were hundreds of ordinary country vicars and parish priests who kept the craft alive in difficult times, introduced it to their humble parishioners, and were among the founder members of bee societies and associations wherever they were formed. The clergy were also in the forefront of the 'humanity to honeybees' movement. And it was the search for a way of keeping bees which did not involve the annual destruction of thousands of colonies which led to the first real advances in hive design.

8 The Development of Modern Methods

Until the Reverend Butler's time people in Britain accepted that bees were kept in skeps and killed with sulphur in the autumn, but he was not the only person to feel that there should be some more humane way. I have heard a ninety-year-old man from the New Forest in Hampshire tell how his father would dig the pits and prepare the sulphur 'matches' and then sit at his kitchen table in the lamplight, nerving himself for the awful deed; how he would go out and return several times before he could bring himself to commit what he obviously felt was a crime. There must have been hundreds of others during preceding centuries who loved their bees and hated the black day when they must be slaughtered; but they knew no other way.

Skeps had many disadvantages—it was difficult to examine the bees or to feed them, and impossible to control the production of queen cells and drones; the space could not be expanded and contracted to meet the needs of the colony except by adding an 'eke' (a sort of topless skep put underneath); the honey was mixed up in the combs with brood in all stages of development, which had to be sacrificed; and skeps were liable to catch fire when the bees were being smoked. Their advantage was that they were cheap. Any handy countryman could make them by the fire on winter evenings and the materials cost nothing. Hives only began to improve when the clergy and others who were less penny-pinched started to take an interest: they could at least afford timber.

In the mid-1600s, the Reverend William Mew of Eastlington in Gloucestershire designed an octagonal wooden hive with additional boxes above, and communicating holes which could be closed with swivelled covers between. This was the beginning of the 'super' or separate honey-box in which the queen did not lay eggs. The story goes that Mrs Mew found the design and had the hive made and set up in the garden while her husband was away in London attending the Westminster Assembly of Divines. Possibly their son took the design with him to Oxford because Dr Wilkins, the Master of Wadham College, had such a hive: Christopher Wren, then at All Souls, sent a sketch of it to Samuel Hartlib

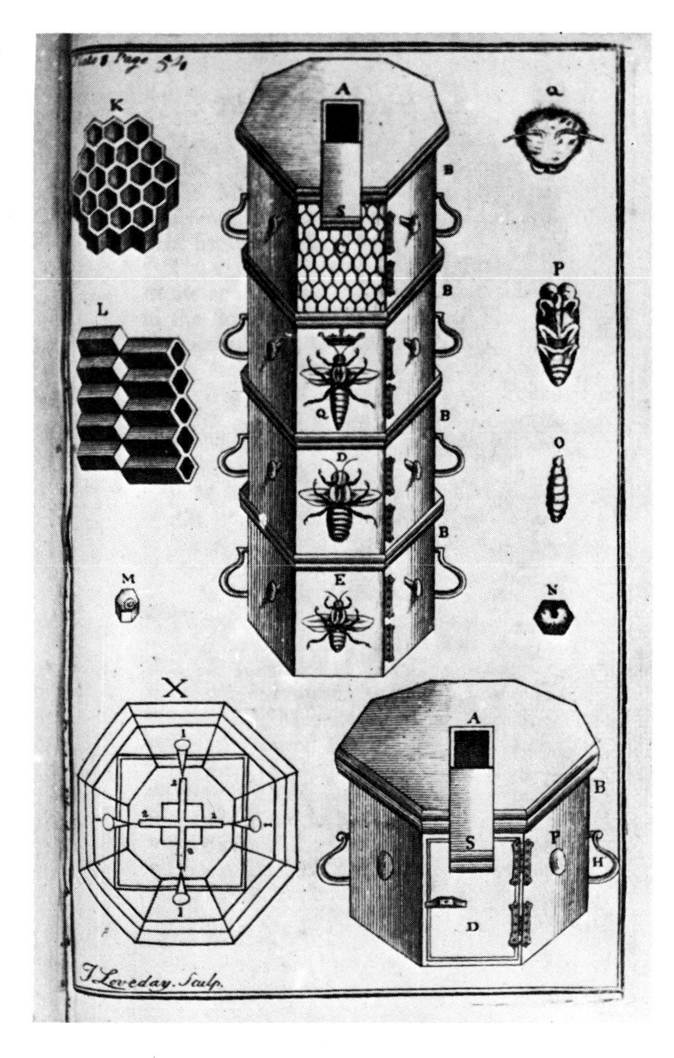

58 John Gedde's wooden beehive with supers (1675)

who printed it in his book *The Reformed Commonwealth of Bees* (1655). This Dr Wilkins was a founder of the Royal Society, which encouraged and publicised scientific research under the patronage of Charles II, who with all his failings was an intelligent man, and appreciated a learned and lively mind as much as he did a pretty woman. In 1675 the King granted a patent to one John Gedde, a Scot from Fife, for his wooden bee-boxes, which he and his partners manufactured and sold for several years. These boxes were very like Mew's hive but contained a removable framework, inside which the bees were supposed to build their combs; though some people objected that they did not do so. The hives were made of thin wood and were intended to be used in a bee-house. Later Moses

Rusden, an apothecary who was also the King's Bee-Master in charge of the hives in St James' park, took over the sale of these hives which he described at length in his book *A Further Discovery of Bees* (1679). He also stated that the Big Bee was a King (possibly as a compliment to his Royal employer).

Charles II, in 1682, knighted a certain George Wheler who had recently spent some time in Greece and published an account of his travels. His description of beekeeping there, and in particular of the hives in use, attracted great interest. These hives were of wicker, in the shape of 'dust-baskets' ie tapering towards the bottom and plastered inside and out with clay. On the top rim rested flat sticks, from which the bees were persuaded to build their combs by fastening small pieces to them as guides. Only the top of the comb sides would be attached to the hive wall. With care this attachment could be cut and the bar with its dependent comb lifted out (Plate 59). This meant that the honeycombs could be detached and the bars replaced for further use; but, too, it was possible to make artificial swarms by moving some of the brood combs to a new basket and filling the spaces in both hives with fresh bars. The top was covered with a roof or hackle of some sort, and the artificial swarm was put in the place of the parent hive, which was moved to a distance. This caused the flying bees to join the new colony (which would raise an emergency queen)

59 The Greek *anastomo cofini* with movable bars was used in the seventeenth century
IBRA Collection

and made it unlikely that the parent one would send out another swarm.

It would be interesting to know when this type of hive first made its appearance in Greece. Had Aristotle had them, he would have been able to examine combs in a way which was impossible with other hives, and his superior knowledge would be explained. However, had they been in use so early, the Romans would surely have adopted them—as they did most Greek ideas. Since none of the Roman authors mentions them, one must assume they were a later invention.

This Greek hive offered a simple way of taking honey without destroying the bees, and it did not call for expensive timber and complicated carpentry. The idea was eventually adapted to straw skeps. Of course, Wheler's book was known to comparatively few people, but gradually the conception of movable bars filtered down through the beekeeping population. In the meantime, various other kinds of wooden hive were tried: some had the honey-chamber underneath, when it was called a nadir; others had honey-chambers at the sides as in the Reverend Stephen White's brainchild, which he described in a book *Collateral Bee-boxes* published in 1756. These ideas did not make much impression; or not immediately. Growing in popularity was the flat-topped skep with a central hole over which was put another skep, or a glass belljar covered with a straw cap, in which the bees could store honey. There was also a remarkable two-part straw hive called the Remunerator and Preserver which was strongly advocated by the first successful beekeepers' association, the Western Apiarian Society of Exeter (Devon) formed in 1798. The Secretary of this body, the Reverend J. Isaac, was the Unitarian minister of Moretonhampstead, an energetic organiser, and author of an excellent little book on beekeeping. Among other things, the Society gave prizes to cottagers who obtained good crops of honey without killing their bees. (The record was 146lb from one hive, but presumably this included the combs.) Unfortunately the association suddenly ceased after ten active years, and no other beekeepers' society was formed in England until the 1890s.

Many books about bees were published in the eighteenth century and important scientific discoveries (which are dealt with in the next chapter) were made in Britain and in Continental Europe. It was the time of the Grand Tour and increased foreign travel, so educated people were more in touch with the ideas current elsewhere. Despite this, mistaken ideas persisted, and practical advances in beekeeping were almost non-existent. Swammerdam established the sole maternity of the queen bee in the seventeenth century, but many beekeepers were not convinced. A Scot called Robert Maxwell, author of *The Practical Beemaster,* was still saying that the

60 Straw skep with eke and stone floor; the belljar would be covered with a straw cap
IBRA Collection

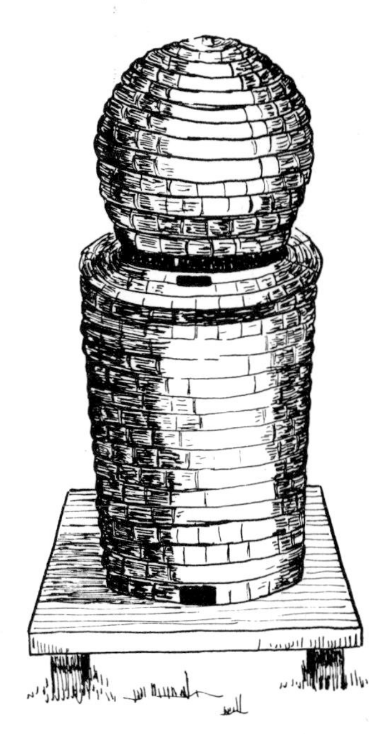

The Remunerator and Preserver hive

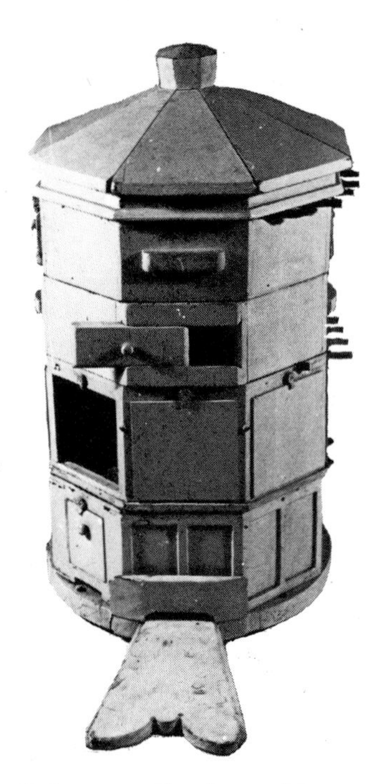

61 Stewarton hive, invented in 1819 *IBRA Collection*

queen was male in 1747; and Daniel Wildman, a man who read widely and who travelled all over Europe giving shows with bees, wrote in 1775: 'I am fully convinced that the modern received opinion that the Queen Bee is the general Parent of their whole stock is absolutely without foundation'; and a certain Mr Allnutt in the *Transactions* of the Western Apiarian Society insisted that queen bees were produced by immuring aged workers in queen cells!

The feeling against killing bees was growing but in general beekeeping went on in the same way as it had in the Middle Ages. John Mortimer FRS in *The Whole Art of Husbandry* (1707) thinks it worth mentioning that 'The general practice now is to cover the face with black Italian crape, such as is used at funerals'. The Reverend John Thorley of Chipping Norton in his *Melisselogia or the Female Monarchy*, amongst his quite advanced anatomical information and masses of pious exhortion, revealed that the smoke from dried puffballs could be used to stupefy bees—by stupefying and uniting two colonies the honey from one could be taken and the bees preserved. The puffball idea was not new: in 1597 John Gerarde mentioned in his *Herball* that 'countrie people use to kill or smoother Bees with Fusseballs being set on fire, for the which purpose it fitly serveth'.

The nineteenth century saw the turning point. In 1819 Robin Kerr of Stewarton in Ayrshire constructed an improved version of the Mew-Gedde type of octagonal hive with supers. It had bars (screwed in place) for the bees to build comb from, and slides between them which pulled out to let the bees up into the next storey. These hives were improved further as time passed. A square version called the Carr-Stewarton did away with bars of different lengths, so making them interchangeable. Later the fixed bars were replaced with close-fitting frames (which the bees glued immovably in place with propolis); and wood and metal queen-excluding strips were substituted for the slides. With modifications, some octagonal Stewartons were still in use at the end of World War II and large crops of honey were obtained from them. This devotion to the octagon seems to have been because it was the nearest approach possible in wood to the round skep (or hollow tree) which the bees were presumed to favour.

At the same time Thomas Nutt was developing the Reverend Stephen White's idea, and in 1832 his book *Humanity to Honeybees* was published. It contained details of Nutt's Collateral hive, which consisted of three boxes placed side by side: the central one was the brood box, and the bees were admitted into the lateral ones to store their honey when more space became necessary. There was also provision for belljars, additional boxes, or even straw skeps, to be added on top; and drawers in the base could be used to supply food for

62 Nutt's Collateral hive, invented in 1832 *IBRA Collection*

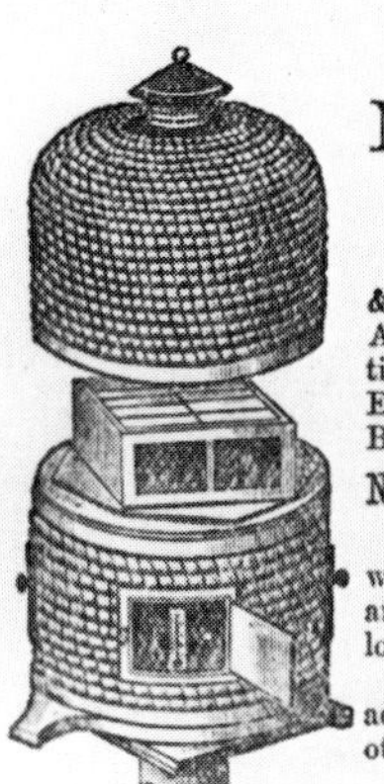

"Gather Honey from Your Flowers."

NEIGHBOUR'S Celebrated BEEHIVES, for taking Honey without the destruction of the Bees.

PHILADELPHIA EXHIBITION, 1876.
PARIS EXHIBITION, 1878.

Three Silver Prize Medals awarded to Geo. Neighbour & Sons. Also the Prize Silver Cup at the Caledonian Apiarian Society's Show at Edinburgh, 1877 (in connection with the Highland and Agricultural Society's Annual Exhibition), for the largest and best display of Bees, Beehives, and Bee Apparatus.

NEIGHBOUR'S Improved COTTAGE BEEHIVE,
as originally introduced by GEO. NEIGHBOUR & SONS,

working three bell-glasses or a tray of sections, is neatly and strongly made of straw; it has three windows in the lower Hive.

This Hive will be found to possess many practical advantages, and is more easy of management than any other Beehive that has been introduced.

Price, Complete, £1 15/; Stand for ditto, 10/6.

63 Advertisement for Neighbour's Improved Cottage Beehive *IBRA Collection*

the bees. Though honey unmixed with brood was usually obtained from Nutt's hives, the boxes were filled with natural comb built down from the top in the same way as the combs in skeps. There are still some Nutt's Collateral hives in existence, although not so far as I know in use. Some are masterpieces of cabinetmaking, occasionally of splendid mahogany, and would have been far beyond the means of a labouring man. Such a person might aspire to the Improved Cottage Beehive sold by George Neighbour and Sons at 127, High Holborn, London (Plate 63). This was a wooden-topped skep with slides, above which stood three belljars or a box of sections covered with a ventilated straw cap; but in 1865 this cost £1 8s, or with shutter-covered glass windows in the lower part, and a thermometer in one, £1 15s. With wages at 10s a week, a farmhand would have been hard put to find that. A cheap commercial or home-made skep, or perhaps a plain wooden box used skep-fashion (the bees being sulphured in the autumn), was still the most common equipment.

The great need and, when it came, the greatest advance in beekeeping, was a hive from which combs could be removed and replaced without damaging them, or the bees. With bars, there was a tendency for the combs to break away as they were lifted, and quite often the bees built diagonally across them. This was to some extent overcome by attaching 'starters' of natural comb, or by glueing slips of wood, triangular in section, along the underside of the bar, as the bees tended to begin their combs there. An ingenious German invention was a metal embossing roller which printed a cross-section pattern of cells in soft wax poured along the bar, to induce the bees to continue in the same line.

The Swiss naturalist Huber had made a hive consisting of separate frames hinged together on one side so that they could be opened out like the pages of a book, and the combs contained in them could be examined on both sides. Although the 'leaf hive' was invented as an aid to scientific research, many attempts were made to adapt it for practical beekeeping. In Huber's hive the frames together formed the actual hive walls, but this was only feasible inside a bee-house. In 1844 William Munn published *A Description of a Bar and Frame Hive* which he had invented. The frames were like Huber's but pushed horizontally into an outer casing; but he also seems to have thought of a top-opening box with hanging frames in 1834, and later, one with quadrant-shaped frames pivoting upwards from one edge of the box. His hive was workable, but like all the other attempts to fit frames into a weatherproof box, the bees propolising them in place rendered them difficult, if not impossible, to remove or separate. All over the world, beekeepers struggled to find a solution to this problem. The Abbé Della Rocca published a book in Paris in 1790 which contained details of a hive with self-spacing (ie

64 Huber's 'leaf hive', which made possible the observation of bees on the combs *F. G. Vernon, IBRA Collection*

shaped) bars. Dr Dzierzon in Carlsmarkt also had bars shaped so that they spaced themselves—they were pushed into his hive from the side and spacing could not be adjusted from above. A Russian called Prokopovich also tried frames which slid in sideways. The American Quaker beekeeper, Moses Quinby, made some which stood on the floor.

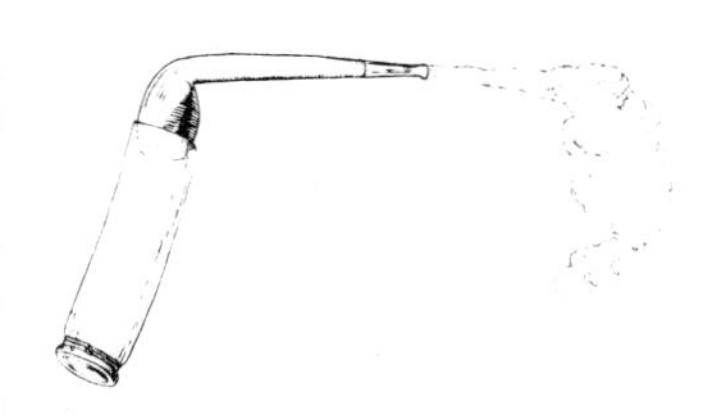

Cheshire's tobacco-pipe smoker

At last in America in 1851 the Reverend Lorenzo Lorraine Langstroth found the answer after many experiments. He made a frame which did not touch the body of the hive anywhere except where the ends of the top bar rested on rebates in the side walls. The bee-space had been discovered; that is, the space which bees will leave clear as a passageway. This is about 6mm or ¼in. Smaller spaces are glued up with propolis; larger ones, the bees fill with comb. It would be nice to report that Langstroth was suitably rewarded for his immense service to beekeeping, but in fact his patent was constantly violated, he was involved in distressing law suits, and eventually gave up the struggle with his health shattered and in great financial difficulties. Later, his cause was taken up by some influential people and he did receive credit for his discovery. He must have been a man of considerable charm, and his book *The Hive and Honeybee* published in 1853 deservedly became a classic.

The spread of the movable-comb hive—the first on sale in Britain was made by the Devonshire beekeeper T. W. Woodbury—completely changed methods of apiculture, and led to a demand for better equipment. A smoker was needed which would produce a controllable stream instead of an asphyxiating cloud, and attempts were made to direct the smoke from rotten rags or touchwood through a tube. Some odd contraptions were shown at the 1874 Crystal Palace Show in London. Frank Cheshire won a prize for a briar pipe fitted with a rubber bulb whereby tobacco smoke could be driven out through the stem, but the amount produced must have been very inadequate. A common bellows with a little burning material in the nozzle was another idea. A metal tube containing fire and kept alight by the beekeeper blowing was tried in America, then in 1875, Moses Quinby attached a bellows alongside a metal tube and the pattern for future smokers was established. Two years later, a development of it, the Bingham smoker, was on sale to the public. Another version, Clark's Cold-blast smoker, was so constructed that air was not driven through the fire, but introduced into the nozzle ahead of it. The cool smoke was supposed to upset the bees less, but the Bingham pattern retained its popularity. Improvements followed: the bellows and tube were separated; a ventilator was incorporated; the nozzle was angled; the firebox enlarged; and the modern smoker was with us.

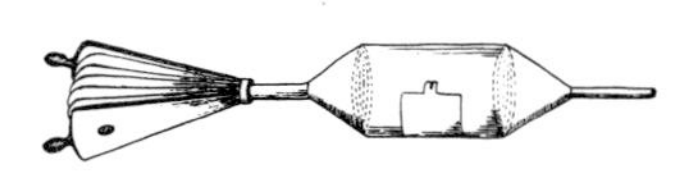

Simple bellows smoker

Wax comb-foundation developed in the same way, as one

65 and 66 Smokers: (top) invented by Moses Quinby in 1875; (bottom) Bingham's version *IBRA Collection*

67 Clark's Cold-blast smoker *IBRA Collection*

person improved on the ideas of another. Movable frames are efficient only if the bees build their combs exactly inside them; wax foundation acts as a guide. A German carpenter called Johannes Mehring in 1857 made a press with wooden plates to print a pattern of worker cell-bases on sheets of wax: the cells had no side walls so drone comb was still built, but the combs were straighter. Improved metal presses were made in Switzerland, Germany, England and America, then A.I. Root of Ohio had a machine built with embossed metal rollers. This was a step forward, but the flat sheets to pass through the rollers were made by dipping wooden boards in tubs of molten wax up to five times—by hand—and peeling off the coating as it cooled. Eventually, in 1896, E. B. Weed of New York State made a machine which delivered a continuous sheet of wax, which was wound on a spool and then fed through rollers, producing at last a reliable commercial product. In England, James Lee and Son obtained from Root's the machinery and the exclusive right to manufacture the commodity, which was called 'British "Weed" foundation'. Nowadays several manufacturers supply foundation, and small presses are available for beekeepers wishing to make their own.

Beeswax was always an expensive material and substitutes, even to aluminium honeycombs, were tried at various times without much success. The invention of an extractor which allowed combs to be emptied and used again meant a huge saving. The story goes that Major von Hrushka, an Austrian, gave his son a piece of honeycomb which he put in a basket. In the way of boys, he whirled the basket around by the handle to drive away the bees buzzing about him, only to find

68 and 69 Modern foundation-making: (right) bonding machine, converting molten wax into a 'ribbon'; (below right) embossing machine, printing cell pattern and cutting sheets to size
L. L. Thorne

70 Close-up of embossing rollers *L. L. Thorne*

the comb emptied of honey by centrifugal force. Langstroth said later that he should have recognised the possibility as he had noticed water being thrown from a grindstone or carriage-wheels many a time. Von Hrushka's first extractor was a metal cone-shaped vessel with a tap at the pointed end and ropes at the top by which it was whirled around. A version of it, the Italian *smielatore,* had a mesh-bottomed box for combs in the top, and was spun round a horizontal pole. Once the principle was understood, dozens of so-called honey-slingers were made, especially in America by Dadant, Quinby, Langstroth and Root. The American Peabody extractor (Plate 71) was patented in 1869: the container as well as the basket holding the combs rotated, and this was not very successful. In England, T. W. Cowan invented a good extractor with a reversible basket so that combs could be turned without being removed, a feature which was soon taken up in America. An earlier idea of Cowan's, an extractor with the frames arranged radially, was later revived when power-driven models were made. For the man with one or two hives, the 'Little Wonder' extractor emptied a single comb at a time, and could even be made by a cottager. The delightful little gadget (see Plate 73) is a honey-press, which forced the honey from a small piece of comb—just enough to fill a pot for the tea-table.

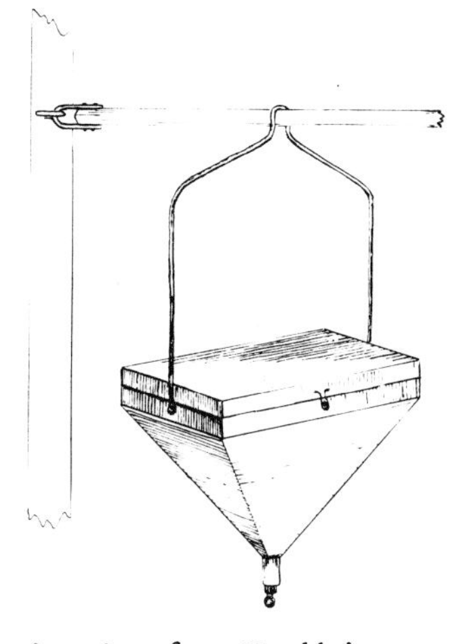

A version of von Hrushka's honey-slinger, the *smielatore*

Being whirled in an extractor put a great strain on the combs, which tended to come out of their frames; and the weight of the bees drawing out the foundation often pulled it away from the top, or distorted the sheets. Various means of supporting them were tried. In America, Dr C. C. Miller inserted five vertical wooden splints—first boiled in wax—in each comb. In England, Frank Cheshire experimented with

71 The American Peabody extractor

72 Abbott's Little Wonder honey-extractor for a single frame *IBRA Collection*

73 A small honey press of the 1890s *IBRA Collection*

metal 'foundation-fixers', like very widely-toothed combs, which were fastened across the wax from top to bottom bars of the frame, but removed when the comb was partly built. Captain Hetherington of Cherry Valley, New York, is generally thought to have originated the use of wires passing through the frame and embedded in the wax to support it. It was not until 1920 that Henry Dadant in America began selling foundation with corrugated vertical wires put in by machinery during manufacture.

The origin of queen-excluders is not easy to determine. Quite early in Britain wooden boards with holes drilled through them had been used to separate brood and super skeps, but not only did small queens pass through them, much of the pollen was scraped off the workers' legs by the round holes. Slotted wooden excluders are said to have been used between the bars of Stewarton hives as early as 1849, but this is difficult to verify. The Abbé Collin in 1865 devised slotted metal sheets, which were widely copied. Tin or zinc was used and the slots were generally about $\frac{9}{50}$ in wide. The sharp burrs round the cut edges tended to injure the bees, but this was obviated in the excluder patented by A. I. Root in 1907 which consisted of galvanised wires passing through flat metal strips. The proper spacing was now precisely defined as $\frac{165}{1000}$ in between the wires.

The last great aid to beekeepers was another American idea: this was a really effective way of clearing bees out of the supers when the crop was taken. E. C. Porter of Illinois in 1891 introduced the bee-escape, apparently designed by his father. This allowed bees to push their way between two flexible strips of metal, which then closed together and prevented their return. The Porter bee-escape is still with us and virtually unchanged, though it is possible that the use of the chemical

Porter bee-escape; the modern form is double-ended

74 Woodbury hive with straw-covered sides *IBRA Collection*

Carr-Stewarton hive with straw-covered sides

benzaldehyde as a quick method of clearing supers may supersede it.

In America, commercial beekeeping was established on a practical and economic footing, though doubtless in remote parts bees were still kept in log bee-gums and wooden boxes. In England, the movable frame and other aids to better beekeeping were accepted, but controversy raged over hives. The Woodbury hive was well designed with top bee-space and holding ten 13½in x 8¾in frames. To make it more acceptable to people who still felt that straw was the proper material for hives, one version had sides covered with a sort of straw matting—and very attractive it was. Carr-Stewartons with straw sides were also made by Lee's of Bagshot. This belief in straw lingered for some time and as late as 1870 a Scottish beekeeper called Pettigrew wrote a book advocating very large straw skeps. After that, hives of all sorts and sizes proliferated: the Raynor Eclectic hive, the Abbott Combination hive, Neighbour's Sandringham hive, the Cottager's, the Wells Two-colony, the Reverend G. Castleden's Rotary hive, Taylor's Non-swarming, Howard's Paragon with fancy roof, and many others. Of major importance were the double-walled hive invented by that great beekeeper (and most versatile man) T. W. Cowan; and the modified version of it designed by William Broughton Carr in 1884 and called the WBC hive. Within a few years the WBC became *the* hive in

75 Pettigrew's hive, a very large straw skep *IBRA Collection*

Britain: white-painted, gabled, standing on short splayed legs, it is the thing which leaps to most people's minds when the word 'beehive' is spoken.

The double-walled hive dominated British beekeeping for a long time and this must be due, at least in part, to the status of the designers. T. W. Cowan was a founder-member of the British Beekepers' Association which began in 1874, and soon achieved tremendous influence. It organised shows and conducted examinations for beekeepers, and when affiliated county associations were set up, people holding BBKA Certificates were employed as County Experts to advise members. In addition to lecturing, examining candidates, and judging at honey shows, Cowan was author of *The British Beekeeper's Guide Book* which went into twenty-five editions and was translated into seven languages. Cowan's own hive and the WBC were discussed in detail in the *Guide*, so it is not surprising that thousands of people never considered any other. William Broughton Carr was also a power in the BBKA, and he and Cowan between them ran the monthly *Beekeepers' Record* and the weekly *British Bee Journal*, answering all calls for advice and spreading the gospel of double-walled hives. In later years Carr took an interest in two young men from Nottingham who had begun to make a name in the bee world; the name then was Herrod. In 1909, they took over the *Record* and the *Journal* from Cowan. The elder brother Joseph

The British Bee Journal

Bee-Keeper's Record & Adviser

REGISTERED AS A NEWSPAPER FOR TRANSMISSION ABROAD.] [ENTERED AT STATIONERS' HALL.

Edited by THOS. WM. COWAN, F.G.S., F.L.S., F.R.M.S., Etc., and W. BROUGHTON CARR.

No. 497. Vol. XIX. N. S. 105.] DECEMBER 31, 1891. [*Published every Thursday, Price 1d.*

76 Heading from an 1891 copy of the *British Bee Journal* showing, left, old straw hives, one with a hackle; Cowan's hive at centre; Stewarton and WBC hives at right

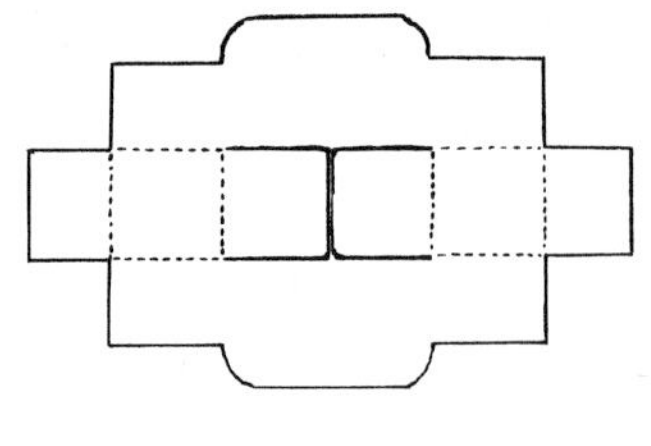

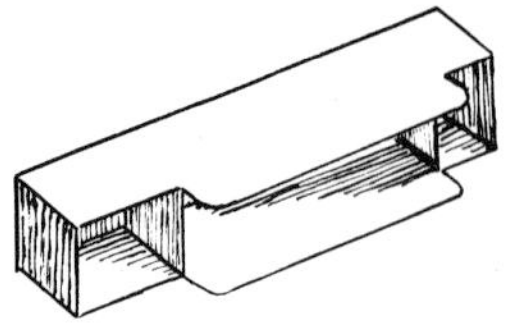

WBC metal-end spacer

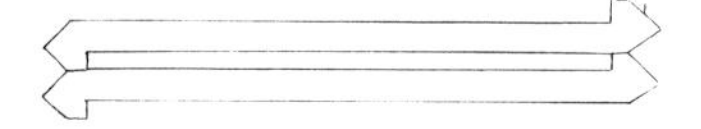

Abbott's self-spacing top bars

ran the peridoicals while the younger, now famous as William Herrod-Hempsall, became the final arbiter in all matters regarding the bee. From a 6d-a-day gardener's boy, he had risen to a position where he was often photographed shaking hands with Royalty, and none questioned his dicta. In a way he was deemed to be as infallible as the ancient authorities like Vergil had been in the Middle Ages. He undoubtedly did a great deal for British beekeeping but he had a rigid antipathy to American ideas and inventions, unless he could prove to his own satisfaction that an Englishman thought of them first. His huge 2-volume *Beekeeping New and Old* is a remarkable work, but the reader must often smile at the author's insularity, dogmatism, and rather endearing snobbery.

This is not to say that the WBC hive was not a great advance on what had gone before. It replaced the Cowan because it was built of thinner wood and could be—and often was—knocked up from packing cases by any handy labourer; and also because W. Broughton Carr invented the 'metal-end' which slipped onto the lugs of the top bar to provide a cheap and foolproof method of spacing frames. Self-spacing frames had been made by Abbott and others, but until the advent of standardisation and advanced woodworking machinery they were not a commercial proposition. The National hive represented an attempt to make a simpler hive on vaguely American lines which would be more suitable for commercial beekeepers and for those who took their hives to the fruit-orchards and heather moors at different seasons. It was designed to take the small British Standard frame, the dimensions of which had been settled by a committee of the BBKA with more reference to the manufacturers' convenience than to the needs of the bee. It is only since World War II that the merits of the American type of large hive have been recognised in Britain, especially where the prolific Italian race of bee is kept. The present cost and world wide shortage of timber can only hasten the demise of double-walled hives. Experiments with fibre glass and other plastics have been made, but problems of condensation and insulation arise.

However much we speculate about the future of beekeeping, we may be sure the bee will not radically alter her way of life. Bees are not tame or domesticated: as long as we make them comfortable they will live in the homes we provide for them, but they do not depend on us as do the stalled ox and the battery hen. It is, in the end, the bee which decides.

9 The Scientists

The first author whose work on the honeybee has been preserved is Aristotle, although it is known from quotations in later books that scholars before his time had written about it. Most of his information about bees can be found in *Historia Animalium,* Book V and *De Generatione Animalium,* Book III. Aristotle was a curious, observant, painstaking man—a true scientist. Given the equipment he had, his conclusions are brilliant. He could only examine a bee with the naked eye, or perhaps some form of primitive magnifying glass, and the hives in use in his day did not permit him to study what went on inside. Removing a comb meant damaging the colony without advancing the research because the bees, suddenly exposed to light, would no longer behave in a normal manner. Bees could be watched when they swarmed or while they were foraging but the secret life in the darkness of the hive remained a mystery. Despite this frustration, Aristotle came close to understanding how bees reproduce. It was 2,000 years before a man of comparable ability applied himself to the problem.

On the whole, few advances in the study of bees have been made by practical beekeepers, at least until modern times. Beekeeping was always the province of humble people without the mental discipline to observe and record, even had they thought it useful to do so. Even nowadays some beekeepers tend to dismiss scientific research as irrelevant, although practical improvements in the craft have always followed—sometimes long after—the discoveries made by scientists. The queen excluder, which guarantees that the honeycombs in supers will be absolutely free of brood, could not have been invented until scientists had proved that the Big Bee was the only one capable of laying eggs.

From the time of Charles Butler, educated men in Britain began taking an interest in the bees themselves, as distinct from the possibility of profit. Such people as the Reverend Samuel Purchas who wrote *A Theatre of Politicall Flying-Insects* in 1657, and the Reverend John Thorley, whose *Melisselogia* appeared nearly a century later, had a more intellectual approach. The former studied bumblebees and wasps as well as honeybees and cut open sparrows which ate his bees, to find them full of drones. Thorley described the eight wax 'plaits' on the bees' abdomen, and discovered for himself that queens were female because one laid eggs as she ran across his hand. Elsewhere in Europe, a similar interest had gradually developed among the educated middle and upper classes.

The two inventions which made the scientific study of bees possible were the microscope and the observation hive. A Dutchman called Anthony van Leeuwenhoek (1632–1723) is generally credited with making the first really workable microscope. Before his time there had been various sorts of magnifying lens and Prince Cesi in Italy had produced some large scale drawings of the bee and its parts in 1625. With this microscope, Leeuwenhoek was able to see and describe things which measured only $\frac{1}{10{,}000}$ in across.

The tool was at hand, and so was the genius able to make full use of it, another Dutchman, Jan Swammerdam, who was born in 1637. He was intended for the Church but instead

77 Jan Swammerdam, who established the maternity of the queen bee in the seventeenth century

took up medicine, although after getting his Doctor of Physic degree he never practised. Instead he returned to the study of natural history which had fascinated him as a boy. His first book was on the *Ephemera* (dayflies), which he recorded in minute detail from egg to short-lived adult. The study of bees was only a small part of his work but his discoveries were impressive.

Some time earlier, a Flemish naturalist called Clutius had asserted that the queen bee was the mother of the hive, but he had not proved it. With the aid of the microscope, Swammerdam found the ovaries and confirmed the theory. He was the first person to dissect bees, and to work out techniques for preserving his specimens from decay. Moreover, he made all his instruments himself; knives and scissors so delicate they could only be sharpened under the microscope, and glass tubes as fine as needles which could be used to extract moisture and air from the vessels in an insect's body and replace them with coloured fluids to define their structure. When he died at the early age of forty-three his instruments and specimens were dispersed and lost, but his marvellous drawings together with the manuscript of his great *Bybel der Natuure* eventually came into the hands of the famous Dutch physician Boerhaave. It was published in Dutch and Latin in 1737, and in English translation about twenty years later.

An obstacle to progress in scientific discovery before this time was lack of communication between people not only in different countries but in different parts of the same country. The Anglo-Saxons for instance called the queen bee *beo-mothor,* and they apparently knew the sex of the workers too since they are referred to as 'victor women' in an ancient charm against swarming. A ninth century incantation to stop bees absconding, used by monks in south-western Germany, begins: *Adjuro te, mater aviorum,* ... (I adjure thee, mother of the bees, ...) Somehow this knowledge disappeared or remained local until Swammerdam proved the sole maternity of the queen. The sex of the workers eluded him till the end. This would not have been so had he come across a book written in 1637 by an English Puritan called Richard Remnant, who had seen the workers' genitals under a glass and knew them to be female. Remnant was a dealer in bees and mead, not a scientist; nevertheless he was an observant and enquiring man.

Times changed. Following the Restoration in 1660, Britain had a king who had, perforce, travelled widely and had a cosmopolitan outlook and broad interests. Under his patronage, the Royal Society fostered all kinds of scientific research and published in its *Philosophical Transactions* papers by foreign as well as British savants, and reviewed important books in many languages. Among others, it published two papers on bees by Leeuwenhoek, who when he died left his microscope and specimens to the Royal Society. (They were later destroyed by fire.) It also published the complete works of the Italian physiologist Marcello Malpighi (1628–1694) the first man to investigate human tissues with a microscope, and discoverer of (among other things) the Malpighian tubules which function as an insect's kidneys. Such contact between scientists of different countries was stimulating: they were able to enlarge on each others' work, and put forward theories to be confirmed or contested by others carrying out experiments independently.

This trend continued and gathered momentum in the eighteenth century. In England in 1712, Dr Joseph Warder of Croydon published a book called *The True Amazons,* in which he described the drone as a male and worker as female, and gave reasons; but he still believed the workers could all lay eggs, and his knowledge of anatomy was shaky. In Paris in the same year, the Italian astronomer/naturalist, Giacomo Maraldi, published his *Observations sur les Abeilles,* which contained much new and correct information on the internal and external structure of bees, and an account of their life. Unfortunately no translation of this was available in England until thirty years later and then only in an abridged form. In 1742 a book called *Microscope Made Easy,* written by Daniel Defoe's son-in-law, Henry Baker, popularised the instrument. British scientists began contributing to the growing body of bee knowledge on an increasing scale. Outstanding among the papers published by the Royal Society was one by Arthur Dobbs in 1750. His main idea—that wax was the faeces of the bee—was hopelessly wrong, but he knew that bees collected pollen from one species of plant at a time, that pollen was the 'male seed' of the flower, and that proper pollination would not result if they were less selective; also he explained the function of the queen's spermatheca, which had baffled Swammerdam. Another brilliant paper on bees was the work of the eminent surgeon, John Hunter. He described clearly how wax was formed, and recorded the dancing of bees (which no one since Aristotle seems to have noticed), though he thought they were running about to shake the wax flakes from their 'pockets'.

If the microscope revolutionised the study of bees in the seventeenth century, the observation hive provided the stimulus for brilliant research in the eighteenth. For centuries people had been looking for a way to see what went on inside the beehive. As early as the first century, Pliny the Elder mentioned a Roman consul who had hives made from the transparent horn used in lanterns, so that he could watch the young bees emerge from their cells. One cannot imagine he was very successful. Tiny windows were put in some of the early wooden hives such as Mew's (see page 94), but the outer combs are always used for storage so at best a restricted view of bees packing or capping honey would be obtained. In 1665 Pepys mentioned in his *Diary* that his friend Mr Evelyn had in his garden 'a hive of bees, so as being hived in glass, you may see the bees making their honey and combs mighty pleasantly'. This was probably Gedde's hive, but even if it had a larger proportion of glass, nothing of the egg-laying or brood-rearing could have been seen. For this, a single comb contained between sheets of glass was necessary, but the person who actually invented the first one remains unknown. Maraldi is said to have found hives like this in the garden of the Royal Observatory when he went to Paris in 1687.

The Seigneur de Réaumur certainly had single-comb observation hives and with them produced a massive body of work on the bee. This is more startling when one realises that he was predominantly a physicist and mathematician, inventor of the Réaumur thermometric scale, and responsible for discoveries in metallurgy connected with steel-making and the tinning of iron. He also surveyed the forests, fossil beds and mines, and reported on the useful arts and manufactures of France for the Academie des Sciences. Though his huge work, *Memoires pour Servir a l'Histoire des Insectes,* was never finished, it could well have represented the life-work of a lesser man. With his glazed hives and microscope, Réaumur subjected the bees to a searching investigation, which occupies two chapters of Volume 5 of the *Memoires.* This volume was published in 1740.

Réaumur confirmed that the queen was the sole mother of the hive, although he also thought she was a ruler in an absolute sense. He marked queens with colours, and introduced strange ones into the hives to see what would happen; he watched the destruction of surplus queens and the massacre of the drones. He saw eggs laid, brood hatch, and larvae being fed by the workers with regurgitated food. The drawings he made of eggs, larvae and nymphs, and also of the sting and mouthparts of the bee are marvellously accurate. He studied temperature regulation in the hive, the construction of the honeycomb in all its mathematical precision, and the collection of pollen. Even the 'bee-bug', the parasitic *Braula coeca,* did not escape his searching enquiry. He could not understand the secretion of wax; and he could not discover how the queen was fertilised, though he did not believe the theory of an *aura seminalis* pervading the hive which had been put forward by Swammerdam.

It was left for the Swiss naturalist, François Huber, to close these gaps. Blind from his teens, Huber used the eyes and the patient hands of his servant, Burnens, to do the practical work among the bees, but his mind devised the experiments and drew conclusions from the results. The 'leaf hive' he designed to help in his work has already been described (see p 100). Among other things, Huber solved the problem of the origin and manufacture of wax, and threw light on questions as diverse as the necessity for pollen in brood-rearing, the nature of propolis, and the ventilation of beehives. In 1760 a German clergyman called Schirach stated that workers could raise queens from worker larvae by feeding them a special diet. This raised a storm of disbelief, and was contested long after Huber had proved that it was indeed so. But above all, Huber was obsessed with the unsolved problem of the queen's *fécondation.* He disproved Swammerdam's idea of the *aura seminalis* and then disposed of the theory, put forward in 1776 by an Englishman called Debraw, that the drone fertilised the eggs after they were laid rather as fish eggs are fertilised. After years of patiently investigating every possibility, Huber realised that mating took place outside the hive. Furthermore, he noticed that if the mating flight were delayed or prevented the queen could still lay eggs, only that they would all hatch into drones; why this was so he could not understand.

In fact, though Huber did solve the problem of the queen's fertilisation, he was not the first to do so. There was at Vienna a man called Anton Janscha who, twenty years before Huber, described in a book how the queen flies out, mates, and returns with the mating sign on her. Janscha was an illiterate peasant from the Austrian province of Carinthia, one of a family of outstanding beekeepers. He won a scholarship to study engraving in Vienna, and while at the Academy there applied for the post of instructor in a school which the Empress Maria Theresa had started to encourage the art of beekeeping. He lectured daily in the Augarten during the summer, and was later appointed the Imperial and Royal Beekeeper. In 1771 he published a book called *A Discourse on the Swarming of Bees* which shows just how advanced a beekeeper he was. A more comprehensive *Complete Guide to Beekeeping* was published in 1775, two years after his sudden death at thirty-nine from typhus.

Janscha was not a scientist and there is no evidence that he ever owned a microscope; moreover he would have found it difficult if not impossible to read the works of people like Swammerdam or Réaumur—even if he did in fact learn to read—since he only spoke Slovene until he went to Vienna. Janscha did not claim to have made any great discoveries (though as a youth he did invent a swarm-catcher) but he certainly knew a good deal about bees which scientists in the rest of the world did not. One can only conjecture that much of this was common knowledge in his native Carinthia, or at least in his family. He used the large horizontal Carniolan hive with removable floor and ends, and slides in the top above which supers could be placed: this probably allowed a more

Taking a swarm

sophisticated method of beekeeping than any other hive of the time. The hives were made to standard dimensions so that they could be piled up in bee-houses. Apart from describing in his book the fertilisation of the virgin queen, he notes the purposeful collection of pollen and the massacre of drones which follow it, and explains in detail the effects of queenlessness, and various methods of swarm-control, queenrearing and introduction. He assumes that everyone knows how to make artificial swarms although at the time it was being hailed as the great new discovery of the German beekeeper, Adam Schirach. In fact Wheler had described in 1682 how the Greek bee-masters made artificial swarms. Janscha discussed the disadvantages as well as the advantages of artificial swarms and outlined several different ways of making them, most of which were only possible with his type of hive. He was an extremely capable beekeeper and an excellent teacher, and it is astonishing that his work was unknown outside his own country. It is a staggering thought that while the great Huber struggled to discover the secret of *la fécondation,* there was an illiterate Austrian peasant who could have told him in two minutes.

The publication in English of Huber's work in 1806 gave a filip to the study of bees. Once again the ecclesiastical writer was well to the fore. Apart from those clergymen who saw in beekeeping a remedy for the ills of the labouring classes, in that it would keep them out of the beerhouses, supplement meagre wages, and point many valuable moral lessons, there were others with a more inquiring attitude. The Reverend William Dunbar, Minister of Applegarth, owned a glazed beehive; his book written for the Naturalist's Library series was mainly scientific and contained a splendid drawing of bees making wax. The Reverend W. Kirby, author with William Spence of a multi-volume *Introduction to Entomology,* was another scientifically minded clergyman. However, the most famous and influential nineteenth century bee book, *The Honeybee,* was written by a doctor, Edward Bevan. It was more or less the standard work for nearly a century, but was superseded by the scientific works of F. R. Cheshire, T. W. Cowan, and Sir John Lubbock, later Lord Avebury. At a more popular level, little books like the Reverend J. G. Wood's *Bees, their Habits, Management and Treatment,* and a number of cheap guides and manuals, made at least some of the new knowledge available to the humblest beekeeper.

The problem which Huber failed to solve, that of the drone-laying queen, was finally elucidated by Johann Dzierzon, a Polish clergyman, in the middle of the nineteenth century. He had been given some golden Italian queens, very unlike his own black bees, and it was the results of crossing these different races which led him to the truth—that drones come from unfertilised eggs and therefore have no genes from a father. Parthenogenesis in alternate generations of aphids had been known and accepted for some time, and Schirach, at least, had some inkling of it in connection with bees, but when Dzierzon's discovery was published a storm of calumny and disbelief burst over his head. His *Theorie und Praxis des Neuen Bienenfreundes* came out in 1848, but no English version appeared until 1882 when a translation of the revised edition called *Rationelle Bienenzucht* was issued under the aegis of C. N. Abbott, editor of the *British Bee Journal.* Despite the violent opposition to his theory, it was scientifically tested, in particular by Professor Theodor von Siebold, and found to be correct.

Though the great questions which people have been asking since the time of Aristotle are now answered, the answers themselves give rise to further questions. The scientist today is unlikely to come up with a great truth which will revolutionise our thinking, in the way that the discoveries of Swammerdam and Dzierzon opened the door to modern selective bee-breeding. A modern scientist is a specialist who may spend years studying one small aspect of his subject, and any new discovery will probably be the cumulative result of work done by several people in different parts of the world. Much about the bee remains obscure, and as has been the case for three hundred years, theories are still put forward to be queried and confirmed or refuted by the research of others. It is beyond my ability even to outline the progress in bee knowledge made in this century: I am no scientist and much of the work is beyond my understanding. Some is even vaguely repugnant to one who regards the bee with affectionate respect rather than clinical interest.

There is however one discovery which seems to me to rank with those of Swammerdam, Schirach, Huber and others: this is Dr Colin Butler's discovery of 'queen substance'. Here at last is the explanation of how the colony is ruled, what the 'spirit of the hive' really is, what gives the signal for swarming or supersedure or for workers to start laying eggs. We have not begun to make use of this knowledge in beekeeping—we have seen how long it takes for the practical applications of scientific discoveries to dawn on practitioners of the craft—but one day, might it not revolutionise the manner in which we control our bees?

10 Beeswax

Nowadays we tend to think that the *raison d'etre* of bees is honey, but this was not always so. To primitive man, the contents of the bees' nest—comb, honey and brood—were food, and he made no attempt to separate them. Later, it became clear that wax had specific useful properties and it was regarded as more important than honey. Although honey was the only sweetener, sugar-cane being unknown beyond the regions where it grew, our ancestors did not have the sweet tooth which we have developed by eating so much sweetened food, so this was of less consequence. Honey was valued as much for medicine as for food, and also as the source of a potent alcoholic liquor. Wax could be melted down, strained and bleached, and made into candles which gave a far better light than anything known before; it could be used for sealing and waterproofing; and being malleable it was easily shaped into representations of people and things. Nothing similar existed, and the demand always tended to exceed the supply.

The source of this versatile material was a mystery for centuries. Bees used it to make their combs, but where did they obtain it? Aristotle believed that they fetched it from flowers and trees on their legs, an obvious confusion with pollen or propolis. This was undisputed for centuries. Charles Butler described how they gathered it 'only between *Taurus* and *Virgo,* unless *Aries* bee warme and kinde, for then they may begin in that month'. He noticed that the workers of a swarm, in their haste to build combs, tended to drop their burdens, and that they lay on the hive stools 'like unto the white scales which fal from yong birds feathers'. He went on to say that some people thought they were in fact scales from young bees' wings, but if they were held in warm fingers they were clearly seen to be wax. He said that people—including Aristotle—had confused pollen and wax, but he also thought both were carried on the bee's legs. Shakespeare too subscribed to this notion since he refers to the bees' 'waxen thighs' which Titania's fairies were to amputate and use as candles.

However, the enquiring minds of the eighteenth century looked more deeply into the matter, and eventually John Hunter and the great Swiss naturalist Huber established that wax is a glandular secretion which appears as flakes in the eight wax pockets on the underside of the bee's abdomen. Chemically speaking it consists of wax esters, wax acids, and hydrocarbons, and is pure white at first, darkening in use. Opinions differ, but the bees probably consume at least 10lb of honey (or sugar syrup) to produce 1lb of comb. This naturally affects the amount of the crop. It has been estimated that a beekeeper who takes 50lb of honey from a colony which built all its comb from scratch, would have had 65lb had he given them foundation, and 80lb if they had received ready-drawn comb. Accurate figures would be difficult to obtain since factors such as the weather, forage available, type of bee and strength of the colony also affect the quantity of honey stored. Also bees occupied with comb-building would not be available to forage, and this would reduce the amount of nectar gathered.

One of the earliest mentions of wax in use is in the story of that ancient Athenian aviator, Daedalus, who made wings from wax and feathers to enable him and his son, Icarus, to escape from Crete. Icarus forgot that a low melting point, around 63°C (146°F), is an essential property of beeswax, and his wings disintegrating as he soared towards the sun, he tumbled into the sea and was drowned.

Wax may not have been an efficient adhesive—though Vergil says Pan made his pipes by 'joining with wax the unequal reeds'—but it made an excellent airtight seal for urns and jars containing foodstuffs, or indeed as the Egyptians

78 Roman wax tablets found in Egypt *British Museum*

discovered, coffins holding the embalmed bodies of the pharoahs. Perhaps someone, thousands of years ago, saw a mouse which bees had coated with propolis to prevent its decay in their nest; he may have mistaken the propolis for wax, copied the idea, and found it worked. The Persians and Syrians both covered the bodies of their important dead with wax before burial. The kind of shroud called a cere-cloth or cerement was a wax-impregnated cloth (*cera* is Latin for wax) which was wrapped tightly around a body to exclude air—a cheap substitute for embalming. Records show that Edward I of England, Henry VII's second daughter, Elizabeth, and Sophia, the infant daughter of James I were buried in them. Cering candles, for waxing cloth, were sold by wax-chandlers until the nineteenth century. The term cere-cloth is also used by Charles Butler in the sense of a poultice or salve, mainly composed of wax, spread on a cloth. He gives directions for making 'a cere-cloth to refresh the wearied sinews and muscles' and another 'to comfort the stomach'.

The Romans found wax-covered wooden tablets an ideal medium for notes and brief letters. The characters were incised with the pointed end of a stylus, the flattened end being used to smooth the wax for re-use or to correct mistakes. The wax was contained in wooden frames, hinged to close like a book, but with the rims raised a little to prevent the surfaces touching. This could be fastened with a blob of wax, perhaps impressed with the sender's device or initials, before a slave delivered it. The Romans sealed legal documents in this way,

79 Ivory stylus for writing on wax. Greek, fifth century BC *British Museum*

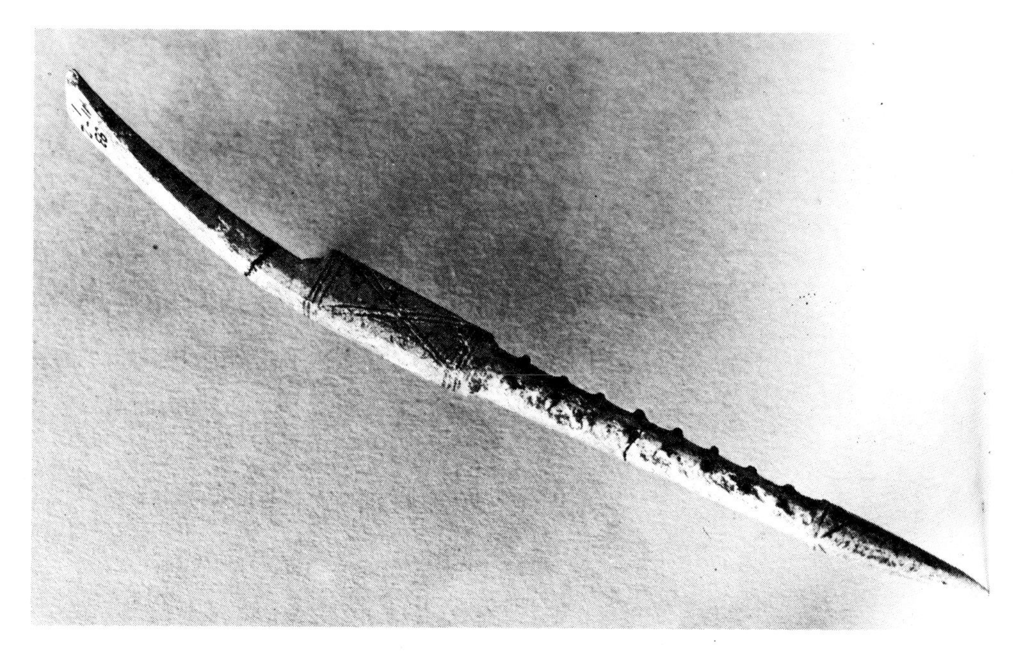

and the official seal has survived in use ever since. Shakespeare in *Henry VI* makes Jack Cade complain, 'Some say the bee stings: but I say 'tis the bee's wax; for I did but seal once to a thing, and I was never mine own man since.' He clearly omitted to read the small print!

Then there were artistic uses for this marvellous material. It was the vehicle for the colours in the technique known as encaustic painting, which was said to last indefinitely. Certainly there are some mummy portraits from the al-Fayyum district of Egypt which have survived in almost perfect condition for upwards of fourteen centuries, though the art itself is much older. As modelling material it was used in many ways. *Cire perdue* is a system used for casting in bronze from very early times. A solid core is made and overlaid with wax which is modelled into the perfect representation of the head or figure to be cast. Over this, more refractory material is applied layer by layer, and supports which pass through the wax connect it to the core. The whole assembly is then heated in an oven until the liquid wax runs out, to be replaced by molten metal. After cooling, the core and outer shell are removed. That erratic genius, Benvenuto Cellini, made his famous statue of *Perseus with the Head of Medusa* this way, and complained bitterly even then of the price he had to pay for beeswax. Present day art students still learn the technique but use micro-crystalline wax (a by-product of oil) which is cheaper.

We think of the Victorians in connection with wax flowers and fruit—and very convincing they were, unlike today's plastic version—but Varro, writing in first century Rome, praised one Posis who could make grapes and apples so realistic that they defied detection. In Rome, too, sculptors made wax busts of their famous men. According to Pliny, Lysistratus of Sicyon made the first accurate (ie not idealised) portrait of a person's face by taking a plaster impression directly from the features and casting a mask in beeswax. In Ancient Egypt, votive statuettes of gods and kings were put in tombs, and similar figures had a religious significance in many other lands. Sculptors like Michelangelo made preliminary models in wax; and portraits modelled in wax became very fashionable in some periods of history. In the sixteenth century, Antonio Abondio was making portrait medallions in Vienna; and a famous French exponent was Antoine Benoist who modelled Louis XIV's courtiers at Versailles, and also visited the English court. He was ennobled by the French king and took as his arms three sable bees on a gold ground. In the eighteenth century, John Flaxman was making wax portraits and reliefs, which Josiah Wedgwood reproduced in pottery.

Life-sized wax effigies preserved in the Norman Undercroft of Westminster Abbey include those of Charles II, Queen

80 Beeswax seal of Robert FitzWalter of Dunmow (died 1235) *British Library Board*

Anne, and Frances, Duchess of Richmond and Lennox (La Belle Stuart) with the stuffed body of a grey parrot which she kept as a pet for forty years. Effigies of the eighteenth century Duchess of Buckingham (an illegitimate daughter of James II and Catherine Sedley) and her sons aged three and nineteen were actually carried at their funerals. The figure of William Pitt was modelled from life by an American, Mrs Patience Wright, reputedly a spy; and that of Nelson was the work of Miss Catherine Andras, 'Modeller in Wax to Queen Charlotte'.

However, the most famous wax models in the world must be those of Madame Tussaud's Exhibition. A Swiss doctor called Philippe Curtius opened a small display of wax figures—his *Cabinet de Cire*—in Paris in 1770. His niece and heir, Marie Grosholtz, showed such an aptitude for the art that she was engaged to teach modelling to the King's sister at Versailles. Came the Revolution, and she found herself making death-masks of the victims of Madame Guillotine (including those of her royal acquaintances), which were kept as a record of the fate of tyrants. Dr Curtius was a whole-hearted supporter of the revolution; Marie did as she was told. Perhaps she recoiled as the bleeding heads piled up in her workshop (a prison cell) but she was tough and kept her own. The death-masks of Louis XVI, Marie Antoinette and Robespierre, looking surprisingly peaceful, are still in the exhibition.

Although she married at the age of thirty-four, Marie lived

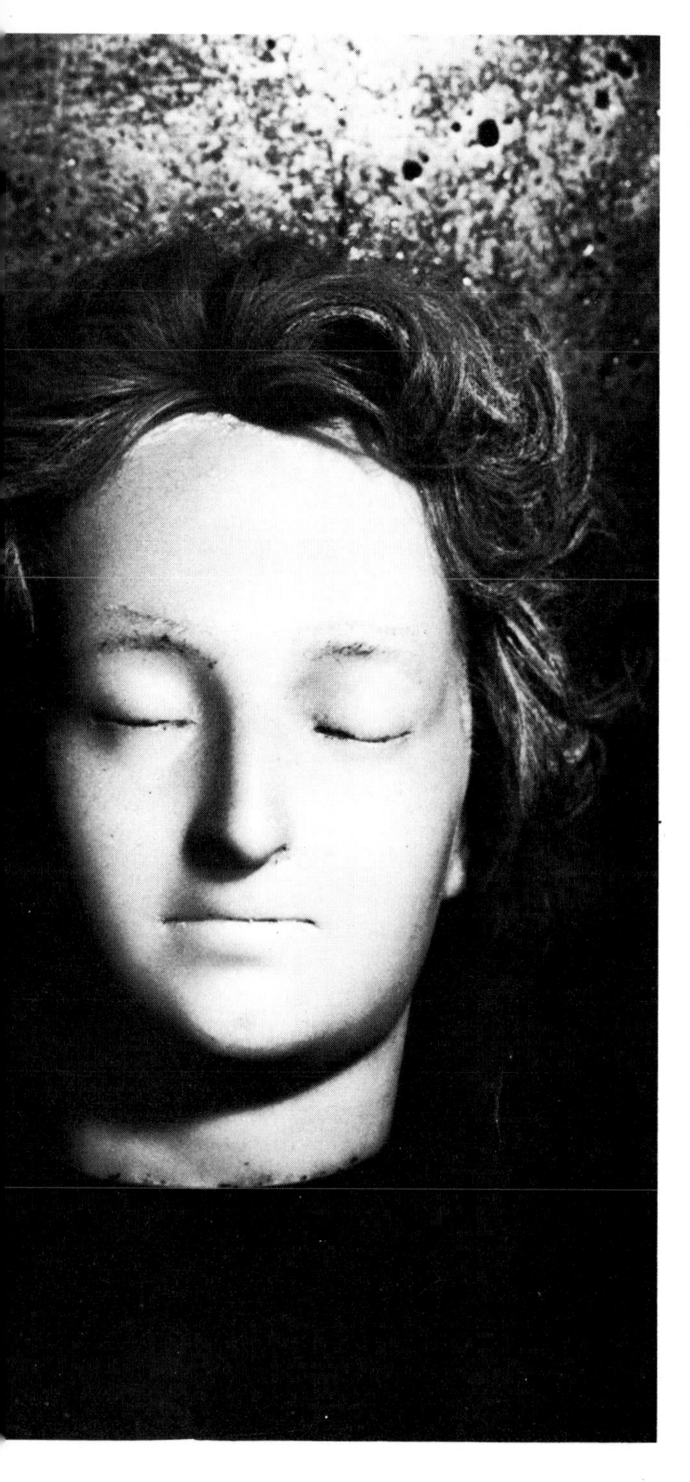

81 Beeswax death masks of Louis XVI and Marie Antoinette, made by Mme Tussaud after their execution *Mme Tussaud's, London*

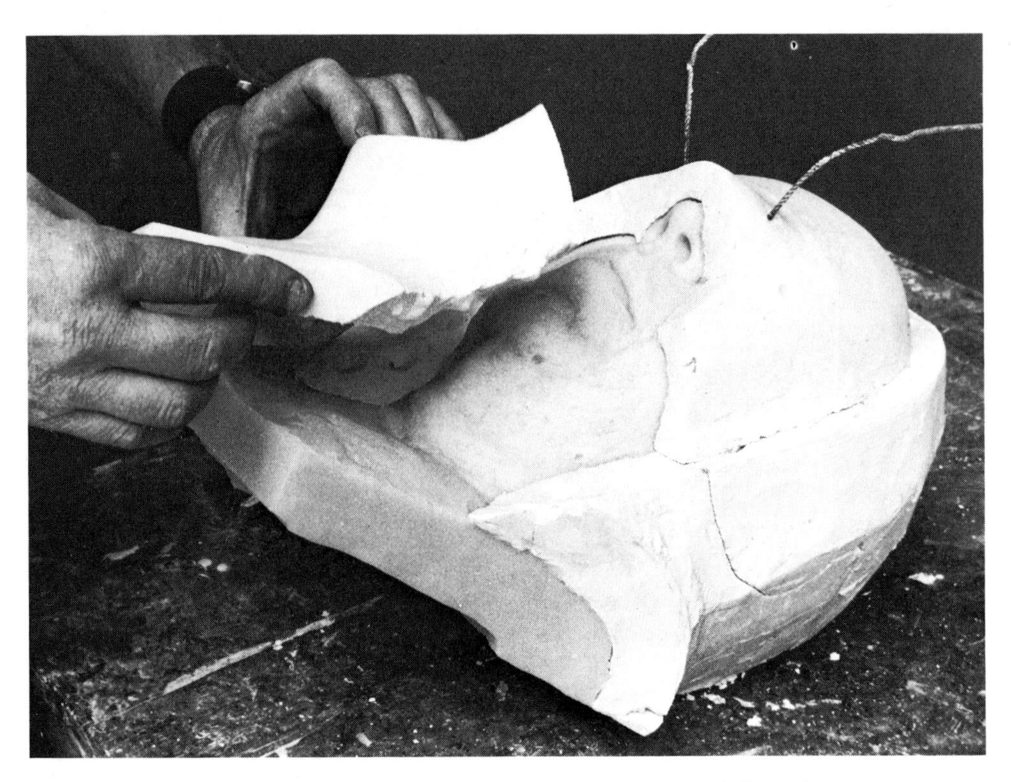

82 Casting a life-size wax head at Mme Tussaud's, London *Mme Tussaud's, London*

with her husband Francois Tussaud for only seven years. In 1802, she took her collection to England and travelled about the country for thirty-three years before settling in London. Four more generations of her family continued her work (the last, Bernard Tussaud, died in 1968) but the show continues, and the models are still made in substantially the same way as they were in her day.

The unclothed parts of the figure, usually head and hands, are made from 75 per cent beeswax with 25 per cent of Japanese wax, a vegetable product. Only wax gives the realistic translucent effect. The figure is sculpted in clay, then a plaster mould is made from the head and tinted molten wax is poured into it. The head is hollow so that a hand can be inserted to fasten the carefully matched glass eyes, and real hairs, about three hundred to the square inch, are put in separately by hand. The final colouring of the face is done with water colours. Concealed parts of the body are cast in fibreglass nowadays, though plaster was used formerly. For two hundred years, this one concern has used beeswax to perpetuate the appearance of the famous and notorious. Even in old age, Mme Tussaud visited the death cells to take particulars of condemned men.

A novelist, Guy Thorne, in 1912 wrote of the 'great sinister dolls' in the Chamber of Horrors, but small and far more sinister dolls had been made from beeswax for centuries before Mme Tussaud was born. These were images fashioned in the likeness of an enemy, often incorporating a lock of his hair or a shred of his clothing, and then stuck with pins or melted on a fire. As the image was injured or destroyed, so the person it represented was supposed to suffer agony or death. Sorcerers and witches in ancient Egypt and India, Greece and Rome, made wax figures for this purpose, and the practice of magic murder was common throughout Europe in medieval times and later. In 1219, Archbishop Gerhard excommunicated the people of Stedding for this offence, and Pope Gregory IX in 1233 threatened all who made such images with damnation. Many cases occurred in Britain: Marjorie Jourdemayne was burned at Smithfield for conspiring with two clerks to kill King Richard III by this means; in 1612 the Lancashire witches, Old Demdike, her daughter Elizabeth Device and eleven year old grand-daughter Alison were accused of it, among other crimes. As recently as the 1940s, a wax doll wearing a scrap of Airforce-blue cloth as a skirt was found in a bombed house with a hatpin thrust through it. Sometimes the figures were baptised, and the potency of the spell was thought to be increased if the image was made from a stolen church candle. On a happier note, some pleasant wax dolls for children were made at various times, but woe betide the little girl who left her treasure too near the fire!

Other uses of wax range from the bizarre to the practical. Ceromancy was the art of divination by dropping melted wax into water: the shapes it assumed as it cooled were supposed to indicate future events. Surviving records show that when the Tudor Palace of Nonsuch was built, beeswax was mixed with resin (wax cost 6½d per lb and resin ½d per 1lb) and used as mortar in damp places where the usual sand/lime mix was unsuitable. There were also medical uses. Wax was esteemed particularly for eye-salves, and as an ingredient in poultices and ointments; 'oyl of wax' was recommended to heal wounds, provoke urine, alleviate pains in the loins and 'other griefes coming of cold'. It was also a recognised cure for dysentery, and it seems that the antibiotic present in wax may be active against certain kinds of entero-bacteria. All the products of the hive contain antibiotic substances (though not the same one). It is interesting to read in Robert Sydserff's *Treatise on Bees* (1792) that bees were themselves regarded as a medicine: he says 'If Bees, when dead, are dried to a powder, and given to either man or beast, this medicine will often give immediate ease in the most excruciating pain, and remove a stoppage in the body, when all other means have failed'. Since it has now been established that the external surface of worker bees (not drones), especially between the seventh and eleventh days after emergence, is covered with an antibiotic substance, perhaps the grisly medicines of the past should not be dismissed as completely foolish. It is true that these antibiotics seem to be most active against bacteria likely to occur in the hive, but it does not rule out the possibility of them being useful to man.

It may be mentioned here that propolis, which the Roman doctors valued beyond wax, is indeed of medical value. This complex composition of resins and balsams, waxes and aromatic oils has been found to possess bactericidal and also fungicidal properties, and in particular is active against certain kinds of tuberculous bacilli. Propolis was found to be an effective dressing for wounds during the Boer War. The Romans used it for treating ulcers and tumours and, according to Pliny, it would draw out stings and other foreign bodies from the flesh. It is at the moment being commercially exploited. In the past it has also been used as incense, especially in Russia, and to make varnish, though the notion that the particular quality of Stradivarius violins was due to the type of propolis available around Cremona does seem to be discounted now.

There is a current belief in some quarters that eating the wax-cappings from honeycomb will prevent or cure hayfever, but this is supposed to be because they contain a good deal of pollen, which taken internally would confer immunity from the effects of wind-borne pollen. It seems to me that the pollens which cause hayfever are unlikely to be the same ones which the bees collect, and I know at least one beekeeper who eats quantities of honeycomb and is nevertheless a martyr to this affliction. The medical uses of pollen have not been thoroughly investigated, but Burmese tribesmen use it to dress wounds with some success, and also make cakes of honey and pollen to carry on hunting expeditions or to store as famine food during the monsoons. Researchers have found that the antibiotic value of pollen from different plants varies considerably, is greater in pollens collected by the bees than in those collected by hand, and tends to increase while it is stored in the cells of the comb. It is rich in vitamins and of high nutritive value, and has been found effective in arresting diarrhoea and relieving chronic constipation.

No extravagant claims are made for wax as a medicine today, but it is non-irritant and a good emulsifier for creams and ointments. Chemists use it as the base for zinc and castor-oil cream, and it is used in considerable quantities by the cosmetic industry for hair-creams, cleansing cream and cold cream. Dentists use pink sheets of blended paraffin wax and beeswax for making impressions for false teeth.

Another early discovery was that wax made a good protective coating for other things beside corpses. Ropes dressed with melted wax lasted longer, and smeared on metalwork it prevented oxidation. It was also used to protect and polish marble. Today the tailor, cobbler and sailmaker still reach for a cake of beeswax to smooth and stiffen the thread used in their work, and beeswax and turpentine still make the best polish for fine furniture. Since domestic servants disappeared, 'easy-clean' wood finishes have become popular, and chemical waxes are used in many furniture polishes. However, beeswax has greater solubility than any other sort and some aerosol polishes contain nothing else. Beeswax is not used in the manufacture of the waxed papers and cartons employed in huge quantities for packaging foodstuffs, though until the advent of commercially produced substitutes, housewives waxed their own coverings for jars of pickles and preserves.

Overwhelmingly the most important use for beeswax until recent times was in candlemaking. As far back as history is recorded, candles played a part in religious observance, among the sun and fire-worshipping Parsees and the Buddhists of Ceylon, as well as in pagan Rome, and in the Eleusinian mysteries in honour of Ceres. Later they became an important part of Christian ritual, particularly at Easter. Taxes and tributes demanded from conquered races were often in the form of wax. When Praetor Pinarius defeated the Corsicans in 181 BC, they had to pay 100,000lb of wax annually, and this was later doubled. Charlemagne taxed the subjugated Saxons in wax, and there are records of Polish princes, German lords, and Tartar kings imposing similar taxes on their subjects. The Christian Church became an enormous consumer of wax. Every monastery and abbey kept bees, but rents and tithes were also paid in wax, and gifts and legacies of wax to ecclesiastical foundations were common.

This was one commodity the humblest beekeeper could sell. He would not have thought of using it himself—the poor went to bed at dusk and being unable to read had no need of a good light. They could manage with the glimmer from a rushlight, a primitive candle made from the pith of a reed dipped in tallow (animal fat, usually mutton). Using wax would have been particularly wasteful in their flimsy hovels; even the palace of Alfred the Great was so draughty that his candles would only burn properly inside lanterns of thin horn. As every schoolboy knows, his clock consisted of candles marked in twelve parts, each lasting twenty minutes, so that six candles covered the twenty-four hours, and he was thus able to devote an equal length of time to the various duties which made up his day. It was a clever idea but only kings and clergy could afford to burn candles in the daylight.

The Feast of Candlemas, on 2 February, was always the day for blessing the church candles. For mass, candles of bleached wax were required, though candles of yellow unbleached wax were used at masses for the dead. It is the size and number of candles used which accounts for the staggering quantities of beeswax consumed by the churches before the Reformation. In ecclesiastical records one reads of a chapel in Einsiedeln where day and night sixteen candles each weighing 30lb were kept burning and of a German church where sixty candles burned on the high altar on feast days. Henry III of England had fifteen hundred candles burning in St Paul's Cathedral on the

83 Beeswax candles made by the dipping method *IBRA Collection*

feast of the Conversion in 1243, and in 1303, candles weighing 1,200lb were burned at the funeral of Richard Gravesend, Bishop of London. There was an enormous secular consumption of candles too. On state occasions, 'Torches of Waxe' were carried in street processions. The torches were large—accounts still exist which mention torches 45in long costing 15 shillings a dozen, and '8 torches using 138lb of wax'—and there might be a thousand for a procession. Some people must have winced at the waste, as people do today when hundreds of pounds worth of fireworks 'go up in smoke' to enliven a celebration. Except in churches and for public occasions, wax candles were only used in quantity by the court and nobility. The household of Thomas, Earl of Lancaster, in 1313 used 1,870lb of wax candles; and also 1,714lb of wax for seals (which alone cost more than £314). Wax candles were used only in the main rooms. Elsewhere, tallow had to suffice. 'Horrible guttering tallow smoked and stunk in passages,' wrote Thackeray, referring to the 'abominable mutton' of his youth. He was born in 1811, which shows how long this state of affairs continued. It is surprising to find that in 1512 John Colet, in the statutes of St Paul's School, ruled that 'In noo tyme in the yere they [the scholars] shall use talought candell in noo wyse but allonly waxcandill at the cost of theyre ffrendes'. One wonders what the 'ffrendes' felt about providing such expensive illumination.

In earlier times, candles were made on the premises by servants or monks. The wicks were repeatedly dipped in a bath of melted wax; or, for larger candles, the wick would be suspended and dippersfull of wax poured over it into a receptacle below, until a sufficient thickness of wax adhered. The wick could be reversed periodically to ensure a uniform thickness. However, the demand was such that craftsmen called wax chandlers gradually took over the trade. A guild of wax chandlers existed in Paris in the thirteenth century and the Wax Chandlers' Company of London was incorporated in 1484, in the reign of Richard III. As with other guilds, the 'freemen of the mistery or craft of wax chandlers' maintained the standards of their trade by regulating the conditions of apprenticeships, and by searching out poor and dishonest workmanship. They also took part in electing city officials, providing men for the city watch, and men and money for the sovereign's wars. The various companies—the wax chandlers were, and still are, twentieth in order of precedence—contributed to the cost of public celebrations such as the triumphal procession following the Armada's defeat, or visits by foreign royalty or great princes of the church, and on these occasions the members lined the streets wearing their livery. The companies were also involved with welfare schemes in the city, such as buying up and storing corn for the relief of the

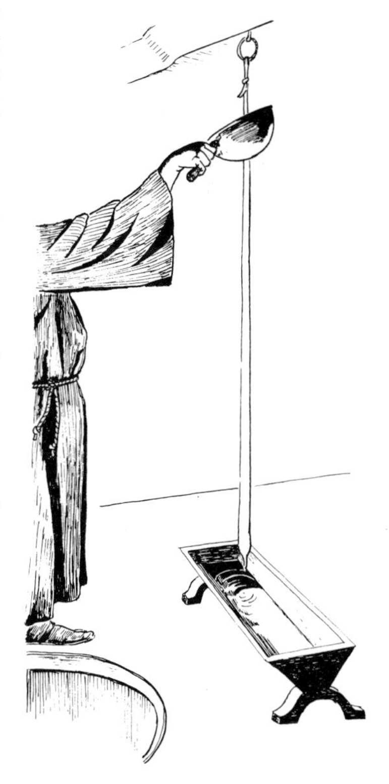

Making a large candle by the pouring method

Beekeeping scenes engraved on the Normansell cup *The Worshipful Company of Wax Chandlers*

poor in times of scarcity. After the suppression of the monasteries, the wax chandlers contributed to the re-establishment of hospitals like St Bartholomew's and St Thomas', which were not originally hospitals in our sense but hostels for the destitute operated by the monks.

The Reformation brought about a great simplification of church ritual. It must have seemed to the Tudor chandlers that almost overnight their best customer disappeared, as the monasteries were closed and candles were forbidden in churches. Prosperous men had been accustomed to leave bequests for daily masses to be said for their souls and lights to be kept burning in chantry chapels in perpetuity but this practice was forbidden as being superstitious and the Crown annexed the money. Roman Catholic ritual returned briefly with the coronation of Mary I, and in the last year of her reign (1558) a Paschal candle weighing nearly 300lb was set up in Westminster Abbey before the shrine of Edward the Confessor. But with her death the candles again disappeared. Earlier, a certain amount of wax was imported from the Baltic through the English east coast ports, but in Elizabeth I's time large quantities of surplus wax were being exported. Much of it was found to be adulterated, so the Queen promulgated an 'Acte for the true melting, making and working of Waxe'. It stipulated that all wares must be made from 'good holsome pur and convenient stuffe', that wax cakes were to be stamped with the maker's initials, and any goods found to be corrupt were to be confiscated and the offender fined.

The power of the guilds gradually declined after the Fire of London, when many tradesmen moved out of the city, but candles continued to be the best form of lighting known. In 1664 Pepys recorded that he had 'begun to burn wax candles in my closet at the office . . . to see whether the smoke offends like that of tallow candles'—before that the Admiralty had always specified best London tallow. In the great houses of the seventeenth and eighteenth centuries, candle-ends were valuable. Candles were never re-lit at the court of Louis XIV of France, and those discarded were worth more to the ladies-in-waiting, whose perquisite they were, than their salaries as members of the royal household. A tall candle was lit and held up while the king prayed before retiring, and was given to some especially favoured gentleman when they left him to sleep. For the coronation banquet of George III, three thousand candles were used to light Westminster Hall. Billheads and tradecards of many eighteenth-century wax chandlers still exist. Some were women, usually widows carrying on a family business, like Eliza Bick who traded 'at the Golden Beehive, opposite the Mansion House, London' in 1758. Her wares included 'white and yellow Wax Candles of all sizes, Branch Lights, Winding and Searing Candle,

Superfine and all inferior sorts of Sealing Wax, Wafers of all Sizes, Flambeaux, yellow and black Links . . .' Flambeaux were the large wax torches used in the streets, and a link was an inferior type, made with resin rather than wax, carried by link-boys escorting sedan chairs. Wafers were discs of wax used to seal a folded letter before envelopes came into use.

The 'closed shop' in candlemaking died out when the industrial revolution replaced master-craftsmen with factory hands, who were not responsible for selling their own product. Today the Worshipful Company of Wax Chandlers plays its part, like the other livery companies, in civic functions and City of London elections. Its wealth, derived from the property it once owned, is used to support medical research, charitable organisations, and educational foundations including the new City University. It retains a link with the ancient craft through various bee associations, and gives an annual prize to the outstanding candidate in the British Beekeepers' Association's senior examination. In 1958 it undertook to supply candles in perpetuity for the new high altar in St Paul's Cathedral—the old one having been destroyed by a bomb in 1940. This continues a connection which has lasted nearly six hundred years, since the time when some of the guild members undertook to provide wax for a candelabrum in the original St Paul's which perished in the Great Fire. The company's greatest treasure is a covered cup of silver presented by Richard Normansell in 1683, and probably specially made since it is engraved on bowl, lid and foot with beekeeping and candlemaking scenes. This, filled with mead, is used as a loving cup when the wax chandlers dine by candle-light in their hall (the fifth on the site), and 'Thy Creature, the Bee' is remembered in their grace before meat.

The use of beeswax declined after the eighteenth century as other waxes became known—spermaceti from whales, stearine which derived from animal fat, and paraffin wax from petroleum. These harder waxes were more suitable for moulding, a method which largely replaced older ways of making candles. The development of paraffin as a clean, safe fuel for lamps, and of gas for lighting, spelt the end of candles, though the light they give is so flattering that they remain popular for dinner parties to this day.

Beeswax candles are still made, and by the old methods, mainly for use in Roman Catholic churches. Originally mass candles had to be 100 per cent beeswax, but this was reduced to 65 per cent in 1850, later to 51 per cent and now—with the current price of beeswax so high, and still rising—to 25 per cent.

Practically all the beeswax used in industry in Britain is imported from the less-developed countries of Africa; and from Australia and New Zealand, where a favourable climate encourages bigger crops of both honey and wax. America buys most of the wax exported from Egypt, South America and Mexico. The Eastern honeybee, *Apis cerana indica,* produces wax of a different chemical composition which will not emulsify so readily and has a limited use. Comparative figures are difficult to obtain, but apart from the small quantities of wax used for purposes as diverse as dental work, sailmaking, and modelling heads, a substantial amount goes into zinc and castor oil cream, a good deal more into church candles and furniture polishes, and a much larger amount into the manufacture of cosmetics, in particular hair preparations. However, the surprising fact which emerges is that the major use of beeswax is in the manufacture of wax foundation for bees, not only in bulk by the hive appliance firms but by hundreds of beekeepers who recycle their old combs.

Alternative materials have been found for manufacturing candles and polishes, but no satisfactory substitute for making honeycomb has been discovered, though many have been tried. Only if foundation is pure beeswax, and stamped with cell bases to the bees' own design and dimensions, will perfect combs be produced from it. Honeycomb is something of an engineering masterpiece, combining the maximum storage

84 Beekeeper recycling old combs: the reclaimed wax will be made into new foundation

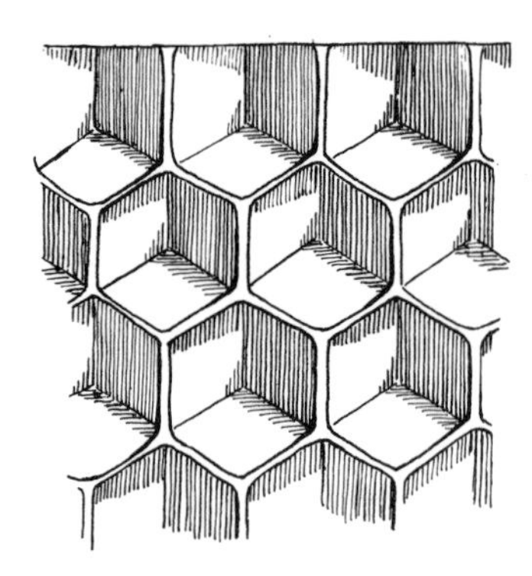

Arrangement of cells in honeycomb

space with the greatest economy of material. The first cells are five-sided to permit a strong attachment to the top bar, and there is a minimum of oddly shaped cells to effect a smooth transition from worker to drone cells, otherwise the hexagonal shape and dimensions do not vary. Worker cells are five to the linear inch (25mm), drone cells are four, and all brood cells are ½in (12.5mm) deep. Honey cells are deeper, but this is a flexible measurement to be varied at need. Nor is the provision of innumerable standard pigeonholes the whole achievement. The cell walls are only $\frac{1}{350}$ in thick, but the strength of this fragile-seeming construction is ensured because each cell base is reinforced by the walls of three cells on the opposite side meeting at its centre, for cells are not built back to back. How effective this is can be gauged if one considers the weight of honey stored in a piece of comb which weighs only ounces, apart from the weight of the bees which constantly move upon its surface. Despite the uses which Man has made of it throughout history, one is forced to the conclusion that the most marvellous, useful and beautiful thing ever made from beeswax is honeycomb.

Throughout the centuries the history of bees has been interwoven with the history of mankind. Man has never really succeeded in domesticating the honeybee—it is as wild today as when it built its nests in the trees of the primeval forests—but it has given him an absorbing field for study and speculation as well as sweetness and light and a drink to cheer his heart, gifts which must always have made his world a pleasanter place in which to live.

Rock-painting from Matopo Hills, Rhodesia, showing bees' nest being smoked *Harold Pager*

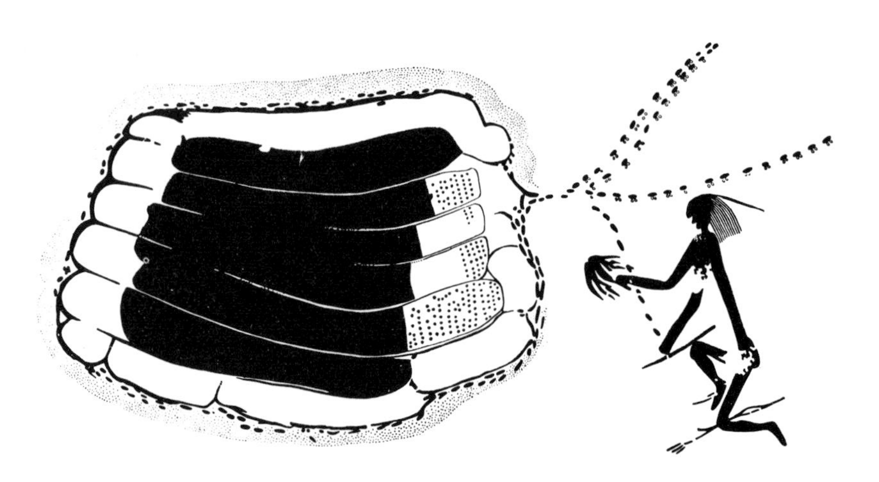

Further Reading

Butler, Colin G. *The World of the Honeybee* (Collins, 1971)

Doering, Harald. *A Bee is Born* (Stirling Publishing Co Inc, 1971)

Dummelow, John. *The Wax Chandlers of London* (Phillimore, 1973)

Fraser, H. Malcolm. *Beekeeping in Antiquity* (University of London Press Ltd, 2nd ed, 1951)

Fraser, H. Malcolm. *History of Beekeeping in Britain* (Bee Research Association, 1958)

Hodges, Dorothy. *The Pollen Loads of the Honeybee* (Bee Research Association, 2nd ed, 1974)

Howes, F. N. *Plants and Beekeeping* (Faber and Faber, 1945)

Nixon, Gilbert. *The World of Bees* (Hutchinson's Nature Library, 1954)

Pellett, Frank C. *History of American Beekeeping* (Collegiate Press Inc, 1938)

Percival, Mary S. *Floral Biology* (Pergamon Press, 1965)

Proctor, Michael, and Yeo, Peter. *The Pollination of Flowers* (Collins, 1973)

Ransome, Hilda. *The Sacred Bee* (Allen and Unwin, 1937)

Ribbands, C. R. *The Behaviour and Social Life of Honeybees* (Bee Research Association, 1953)

Vergil. *Georgics*, Book IV

Index

(Numbers in italic refer to illustrations)